Communications in Computer and Information Science 588

Commenced Publication in 2007
Founding and Former Series Editors:
Alfredo Cuzzocrea, Dominik Ślęzak, and Xiaokang Yang

More information about this series at http://www.springer.com/series/7899

Diana Trandabăţ · Daniela Gîfu (Eds.)

Linguistic Linked Open Data

12th EUROLAN 2015 Summer School
and RUMOUR 2015 Workshop
Sibiu, Romania, July 13–25, 2015
Revised Selected Papers

 Springer

Editors
Diana Trandabăţ
University Alexandru Ioan Cuza
Iaşi
Romania

Daniela Gîfu
University Alexandru Ioan Cuza
Iaşi
Romania

ISSN 1865-0929 ISSN 1865-0937 (electronic)
Communications in Computer and Information Science
ISBN 978-3-319-32941-3 ISBN 978-3-319-32942-0 (eBook)
DOI 10.1007/978-3-319-32942-0

Library of Congress Control Number: 2016936440

Printed on acid-free paper

This Springer imprint is published by Springer Nature
The registered company is Springer International Publishing AG Switzerland

Preface

Social media is a constant in our life, influencing the way we think, interact, learn, consolidate relationships, and understand society. As a result of the rapid worldwide acceptance and usage of social media, more and more content is becoming available as each day passes. Both because of its importance and its increasing volume, it is not surprising that information from social media is rapidly becoming an essential source for natural language processing (NLP) research.

At the same time, linked data is emerging as an increasingly important topic for NLP. Work in the field has produced massive amounts of linguistic data, including annotated corpora, lexicons, databases, and ontologies, in formats that enable their exploitation in the Semantic Web. Linking the contents of these resources to each other as well as to common ontologies can enable access to and discovery of detailed linguistic information and could foster a major leap forward in NLP research and development.

The Workshop on Social Media and the Web of Linked Data, RUMOUR-2015, held on July 18 at Sibiu, Romania, aimed to gather innovative approaches for the exploitation of social media using Semantic Web technologies and linked data by bringing together research on Semantic Web, linked data, and social sciences. The workshop gathered practitioners, researchers, and scholars to share examples, cases, theories, and analyses of social media and linked data in order to address the intersection among these areas. This intersection included not only challenges such as the understanding of and acting upon large-scale data of different kinds, provenance, and reliability, but also the use of these media for crisis management, involving issues of credibility, accountability, trustworthiness, privacy, authenticity, and provision of provenance information.

For the RUMOUR-2015 workshop, we received 21 submissions through the EasyChair submission platform, out of which 10 were accepted for presentation at the workshop and publication in these proceedings. Each of the submitted papers was thoroughly reviewed by three Program Committee members, experts in the topics of the workshop. All papers emphasize innovative approaches for exploiting social media using Semantic Web technologies and linked data, addressing the following topics:

- Ontological modeling of social media data

  - "Ontological Modeling of Rumors" by Thierry Declerck, Petya Osenova, Georgi Georgiev, and Piroska Lendvai
  - "Toward Creating an Ontology of Social Media Texts" by Andreea Macovei, Oana-Maria Gagea, and Diana Trandabăț

– Application of social media and linked data methodologies in real-life scenarios

- "Toward Social Data Analytics for Smart Tourism: A Network Science Perspective" by Mihăiță Antonie, Costin Bădică and Alex Becheru
- "A Mixed Approach in Recognizing Geographical Entities in Texts" by Dan Cristea, Daniela Gîfu, Mihai Niculiță, Ionuț Pistol, and Daniel Sfirnaciuc

– User profiling and assessing the suitability of content from social media

- "Image and User Profile-Based Recommendation System" by Cristina Şerban, Alboaie Lenuța, and Adrian Iftene

– Extracting and linking content

- "Discovering Semantic Relations Within Nominals" by Mihaela Colhon, Dan Cristea, and Daniela Gîfu
- "Quality Improvement Based on Big Data Analysis" by Radu Adrian Ciora, Carmen Mihaela Simion, and Marius Cioca
- "Romanian Dictionaries. Projects of Digitization and Linked Data" by Mădălin Ionel Pătraşcu, Gabriela Haja, Marius Radu Clim, and Elena Isabelle Tamba

– Sentiment analysis in social media and linked data

- "Extracting Features from Social Media Networks Using Semantics" by Marius Cioca, Cosmin Cioranu, and Radu Adrian Ciora

– Social data mining to create structured social media resources

- "Including Social Media – A Very Dynamic Style, in the Corpora for Processing Romanian Language" by Perez Cenel Augusto, Mărănduc Cătălina, and Simionescu Radu

The workshop was a satellite event of the two-week-long EUROLAN-2015 Summer School, the 12th in the series of EUROLAN schools. The traditional organizers of EUROLAN Summer School, the Faculty of Computer Science of Alexandru Ioan Cuza University of Iaşi, Romania, along with the Romanian Academy through the Research Institute for Artificial Intelligence and with Vassar College, New York, USA, were associated, in every edition, with co-organizers holding an impressive international reputation from France, The Netherlands, Belgium, Switzerland, Germany, USA, Italy, Malta, and Portugal.

Each edition of the school has gathered renowned invited professors, interested in this summer school because of its reputation but also because of the proposed topics, which have always been up to date, for example: Language and Logic (1993); Language and Perception: Representations and Processes (1995); Corpus Linguistics and the Awareness on Language Technology (1997); Lexical Semantics and Multilinguality (1999); Creation and Exploitation of Annotated Language Resources (2001); Semantic Web and Language Technology. Its Potential and Practicalities (2003); Multilingual WEB: Resources, Technologies, and Prospects (2005); Semantics, Opinion and Sentiment in Text (2007), and Natural Language Processing Goes Industrial (2011).

This year's edition provided a comprehensive overview of "Linguistic Linked Open Data," including an introduction to formalisms for representing linguistic resources, extracting and integrating knowledge from text, semi-structured and badly structured data, ontologies and reasoning, exploitation of big data using Semantic Web query languages, reasoning capabilities, and much more. During EUROLAN 2015, 12 invited lecturers offered hours of interactive tutorials and hands-on sessions, a positive aspect for an active learning approach. Short abstracts of these tutorials are listed at the beginning of this volume, in the order of presentation during EUROLAN 2015:

- Gerard de Melo (Tsinghua University, Beijing, China): Information Extraction and Knowledge Integration
- Roberto Navigli (Sapienza University of Rome, Italy): BabelNet 3.0: A Core for Linguistic Linked Data and NLP
- Nancy Ide (Vassar College, New York, USA): Annotations as Linked Data – Interoperability
- Steve Cassidy (Macquarie University, Sydney, Australia): The Alveo Virtual Lab: Working with an API for Linguistic Data
- Assuncion Gomez-Perez, Jorge Gracia and Daniel Vila-Suero (Universidad Politécnica de Madrid, Spain): Multilingual Linked Data Generation from Language Resources
- Christian Chiarcos (Goethe University, Frankfurt am Main, Germany): Annotation Interoperability
- John McCrae (University of Bielefeld, Germany): Linked Data for Lexicons
- Thierry Declerck(Austrian Center for Digital Humanities and DFKI GmbH, Austria Multilingual Technologies Lab, Germany): Building Ontolex Resources Using Protégé
- Michael Zock (Université Aix-Marseille, France): Needles in a Haystack and How to Find Them. Building a Resource to Help Authors (Speakers/Writers) to Overcome the Tip of the Tongue Problem
- Dan Cristea and Liviu-Andrei Scutelnicu (Alexandru Ioan Cuza University of Iaşi, Romania): Towards Interoperability of Annotation. Use Cases of Corpora and Software
- Dan Tufiş (Romanian Academy, Bucharest, Romania): Challenges in Building a Publicly Available Reference Corpus
- Tiberiu Boroş (Romanian Academy, Bucharest, Romania): Practical Session on Text-to-Speech Alignment

In addition to the excellence of its academic program, the EUROLAN sequence of summer schools is well known for the camaraderie among professors and students, who may enjoy common activities and social events during dinners and later in the evenings, as well as an excursion in the middle school weekend. The venue of RUMOUR and EUROLAN 2015, Sibiu, was the European capital of culture in 2007, exhibiting an attractive and stimulating atmosphere. With its rich range of restaurants, pavement cafes, and beer gardens, Sibiu is a city inspiring social and scientific networking.

Many people contributed to the success of this event, and we express our sincere thanks to all of them. The Program Committee, made up of international experts in the area of social media and linked data, spent long hours carefully reviewing all the paper proposals submitted to RUMOUR 2015 to ensure a qualitative improvement of the papers. The members of the Organizing Committee enthusiastically assured the settings were appropriate for scientific networking. The invited speakers of EUROLAN 2015, the authors of the papers submitted to the RUMOUR Workshop, along with all the participants, together contributed to transform this event into a warm environment where practitioners and students shared information, ideas, and future research topics. We hope they all enjoyed the technical program as well as the social events of the 12th EUROLAN Summer School and its satellite RUMOUR Workshop.

October 2015

Diana Trandabăț
Daniela Gîfu

Organization

Program Committee

Steve Cassidy	Macquarie University, Sydney
Dan Cristea	Alexandru Ioan Cuza University of Iași, Romania
Gerard de Melo	Tsinghua University, Beijing
Thierry Declerck	Universität des Saarlandes, Saarbrücken, Germany
Daniela Gîfu	Alexandru Ioan Cuza University of Iași, Romania
Nancy Ide	Vassar College, USA
Adrian Iftene	Alexandru Ioan Cuza University of Iași, Romania
Radu Ion	Research Institute for Artificial Intelligence, Romanian Academy, Romania
Vivi Năstase	Fondazione Bruno Kessler, Trento, Italy
Andrei Olariu	University of Bucharest, Romania
Diana Trandabăț	Alexandru Ioan Cuza University of Iași, Romania
Dan Tufiș	Romanian Academy, Research Institute for Artificial Intelligence Mihai Drăgănescu, Romania
Gabriela Vulcu	Insight, Centre for Data Analytics at National University of Ireland, Ireland
Michael Zock	Aix-Marseille Université, France
Dan Ștefănescu	Vantage Labs, USA

Organizing Committee

Anca Diana Bibiri	Alexandru Ioan Cuza University of Iași, Romania
Dan Cristea	Alexandru Ioan Cuza University of Iași, Romania
Daniela Gîfu	Alexandru Ioan Cuza University of Iași, Romania
Radu Ion	Research Institute for Artificial Intelligence, Romanian Academy, Romania
Dan Ștefănescu	Vantage Labs, USA
Diana Trandabăț	Alexandru Ioan Cuza University of Iași, Romania
Dan Tufiș	Romanian Academy Research Institute for Artificial Intelligence Mihai Drăgănescu, Romania
Gabriela Vulcu	Insight, Centre for Data Analytics at National University of Ireland, Ireland

Invited Papers

Information Extraction and Knowledge Integration (Invited Tutorial)

Gerard de Melo

Tsinghua University
gdm@demelo.org

Abstract. The future of computation will crucially depend on our ability to draw on the unprecedented availability of Big Data to produce more intelligent systems that make more informed decisions. There are two major strategies for turning Big Data into knowledge. The first is to extract information from very large amounts of text, relying on pattern-based extraction and mining algorithms. The second strategy is to merge and refine information, often from pre-existing structured sources. This comes with its own set of challenges, entailing a need for special knowledge integration algorithms. Applying these, we can obtain large multilingual databases such as Lexvo.org and UWN/MENTA.

Keywords: Knowledge bases · Information extraction · Data integration

1 From Big Data to Knowledge

In our quest to make information systems better serve our needs, it appears that significant amounts of knowledge and common sense need to be imparted to enable more adequate semantic analyses and more informed decisions. Fortunately, we can now draw on unprecedented amounts of Big Data for this task. Depending on the nature of the data, there are two broad strategies for dealing with it: information extraction (broadly conceived) and knowledge integration.

In the case of unstructured or semi-structured data, our first goal may be to identify meaningful patterns in the data. This can take the form of linguistic pattern matching, as in conventional information extraction, narrowly conceived, and perhaps best exemplified in our research on Web-scale extraction from n-gram data [8]. It can also take the form of exploiting textual signals in machine learning algorithms that automatically perceive meaningful patterns with respect to some goal, such as hypernym acquisition and taxonomy induction [1]. Subsequently, the extracted knowledge can be used in numerous different ways. For instance, we have shown that improved word2vec-style distributed vector representations of words can be acquired if explicit pattern matches bear additional weight during training [3].

The second major challenge is that of knowledge integration. In the case of extracted data, this can be a subsequent step, leading to important refinements. This is the case in our WebChild project [7], which aggregates and disambiguates large amounts of extractions from the Web in order to acquire human-style common-sense knowledge, for instance, that salad is edible and that dogs are capable of barking.

In many other cases, data may already be given in a (semi-)structured form, e.g. as metadata embedded into Web pages or even as curated Linked Data. Still, challenges such as heterogeneous schemata or incomplete and at times even inaccurate links mean that such data is not necessarily readily usable in applications. Instead, additional algorithms need to be invoked to connect equivalent entities [2], to resolve inconsistency problems stemming from wrongly identified entities [5], to produce a coherent taxonomy [5], and to harmonize different schemas, as in the FrameBase project [6], which relies on semantic frames for this purpose. Only then can one arrive at large integrated knowledge sources such as Lexvo.org [4], an important hub in the Web of Data, and UWN/MENTA [5], a large multilingual knowledge graph describing millions of names and words in over 200 languages in a semantic hierarchy.

By smartly applying innovative algorithms to Big Data, we can thus produce new assets that can finally serve us in a number of different ways, be it through better Web search, data organization, or multilingual support, to name but a few.

2 Short Biography

Gerard de Melo is an Assistant Professor at Tsinghua University, where he is heading the Web Mining and Language Technology Group. He has published over 50 research papers in these areas, being awarded, among others, the Best Paper Award at CIKM 2010, Best Demonstration Award at WWW 2011, and Best Paper Honorable Mention at ACL 2014. Previously, de Melo had spent two years as a Visiting Scholar at UC Berkeley, working in the ICSI AI group, as well as 4 months at Microsoft Research Cambridge in the UK. He received his doctoral degree at the Max Planck Institute for Informatics in Germany. Further information is available at http://gerard.demelo.org.

References

1. Bansal, M., Burkett, D., de Melo, G., Klein, D.: Structured learning for taxonomy induction with belief propagation. In: Proceedings of ACL 2014 (2014)
2. Böhm, C., de Melo, G., Naumann, F., Weikum, G.: LINDA: distributed web-of-data-scale entity matching. In: Proceedings of CIKM 2012 (2012)
3. Chen, J., Tandon, N., de Melo, G.: Neural word embeddings from large-scale commonsense knowledge. In: Proceedings of WI 2015 (2015)
4. de Melo, G.: Lexvo.org: language-related information for the linguistic linked data cloud. Semantic Web 6:4. IOS Press (2015)
5. de Melo, G., Weikum, G.: Taxonomic data integration from multilingual wikipedia editions. Knowl. Inf. Syst. **39**(1), 1–39 (2014)
6. Rouces, J., de Melo, G., Hose, K.: FrameBase: representing N-ary relations using semantic frames. In: Proceedings of ESWC 2015 (2015)
7. Tandon, N., de Melo, G., Suchanek, F., Weikum, G.: WebChild: harvesting and organizing commonsense knowledge from the web. In: Proceedings of WSDM 2014 (2014)
8. Tandon, N., de Melo, G., Weikum, G.: Deriving a web-scale common sense fact data-base. In: Proceedings of AAAI 2011 (2011)

BabelNet 3.0: A Core for Linguistic Linked Data and NLP

Roberto Navigli

Viale Regina Elena, 295, Rome, Italy
navigli@di.uniroma1.it

Abstract. This tutorial will introduce BabelNet 3.0, the largest multilingual encyclopedic dictionary and semantic network, which covers 271 languages. BabelNet is a core component of the Linked Open Data cloud and a powerful engine for virtually any Natural Language Processing task in desperate need of wide-coverage lexical semantics in arbitrary languages.

Keywords: BabelNet · Multilinguality · Semantic networks · WordNet · LLOD · Linked Data · Linked Open Data · Knowledge acquisition

1 Outline of the Tutorial

Multilinguality is a key feature of today's world, as is also reflected on the Web, and it is this feature that we leverage in the BabelNet project (http://babelnet.org) which we will overview and showcase in this tutorial. BabelNet [1] 3.0 is the largest multilingual encyclopedic dictionary and semantic network, which covers 271 languages and provides both lexicographic and encyclopedic knowledge for all the open-class parts of speech, thanks to the seamless integration of WordNet, Wikipedia, Wiktionary, OmegaWiki, Wikidata and the Open Multilingual WordNet.

At the heart of BabelNet lies the use of semi-structured, collaboratively-created resources like Wikipedia, from which much-needed knowledge can be harvested [2] and linked to WordNet. Moreover, in order to increase lexical coverage in several languages, additional lexicalizations are obtained as a result of the application of a state-of-the-art machine translation system to sense annotated sentences.

BabelNet 3.0 also integrates the Wikipedia Bitaxonomy [3], therefore providing a full taxonomization of both concepts and named entities. With its 13.8 million synsets and 2 billion RDF triples, BabelNet is a core component of the Linked Open Data cloud and a powerful engine for virtually any Natural Language Processing task in desperate need of wide-coverage lexical semantics in arbitrary languages. The tutorial also included a practical session showing how to query BabelNet 3.0 in Java and via SPARQL [4]. BabelNet is applied to several tasks: not only for the creation of gold standard datasets in Multilingual Word Sense Disambiguation (WSD) [5, 6], but also for the development of joint approaches to multilingual WSD and Entity Linking [7], explicit and implicit semantic representations [8, 9], video games with a purpose [10], etc.

2 Short Biography

Roberto Navigli is an Associate Professor in the Department of Computer Science of the Sapienza University of Rome. He was awarded the Marco Cadoli 2007 AI*IA Prize for the best doctoral thesis in Artificial Intelligence and the Marco Somalvico 2013 AI*IA Prize for the best young researcher in AI. He is the first Italian recipient of an ERC Starting Grant in computer science and informatics on multilingual word sense disambiguation (2011–2016), a co-PI of a Google Focused Research Award on Natural Language Understanding and a partner of the LIDER EU project.

His research lies in the field of Natural Language Processing (including multilingual word sense disambiguation and induction, multilingual entity linking, large-scale knowledge acquisition, ontology learning from scratch, open information extraction and relation extraction).

He has served as an area chair of ACL, WWW, and *SEM, and a senior program committee member of IJCAI. Currently he is an Associate Editor of the Artificial Intelligence Journal, a member of the editorial board of the Journal of Natural Language Engineering, a guest editor of the Journal of Web Semantics, and a former editorial board member of Computational Linguistics.

References

1. Navigli, R., Ponzetto, S.: BabelNet: the automatic construction, evaluation and application of a wide-coverage multilingual semantic network. Artif. Intell. **193**, 217–250 (2012)
2. Hovy, E., Navigli, R., Ponzetto, S.: Collaboratively built semi-structured content and artificial intelligence: the story so far. Artif. Intell. **194**, 2–27 (2013)
3. Flati, T., Vannella, D., Pasini, T., Navigli, R.: Two is bigger (and better) than one: the Wikipedia bitaxonomy project. In: Proceedings of ACL 2014, Baltimore, USA, pp. 945–955 (2014)
4. Ehrmann, M., Cecconi, F., Vannella, D., McCrae, J., Cimiano, P., Navigli, R.: Representing multilingual data as linked data: the case of BabelNet 2.0. In: Proceedings of LREC 2014, Reykjavik, Iceland, 26–31 May 2014
5. Moro, A., Navigli, R.: SemEval-2015 task 13: multilingual all-words sense disambiguation and entity linking. In: Proceedings of SemEval 2015, Atlanta, Georgia, pp. 288–297 (2015)
6. Navigli, R., Jurgens, D., Vannella, D.: SemEval-2013 task 12: multilingual word sense disambiguation. In: Proceedings of SemEval 2013, pp. 222–231, Atlanta, Georgia, 14–15 June 2013
7. Moro, A., Raganato, A., Navigli, R.: Entity linking meets word sense disambiguation: a unified approach. Trans. ACL (TACL) **2**, 231–244 (2014)
8. Iacobacci, I., Pilehvar, M.T., Navigli, R.: SensEmbed: learning sense embeddings for word and relational similarity. In: Proceedings of ACL 2015, Beijing, China, pp. 95–105 (2015)
9. Camacho-Collados, J., Pilehvar, M.T., Navigli, R.: A unified multilingual semantic representation of concepts. In: Proceedings of ACL 2015, Beijing, China, pp. 741–751 (2015)
10. Jurgens D., Navigli, R.: It's all fun and games until someone annotates: video games with a purpose for linguistic annotation. Trans. Assoc. Comput. Linguist. (TACL) **2**, 449–464 (2014)

Annotations as Linked Data – Interoperability (Invited Tutorial)

Nancy Ide

Vassar College, New York, USA
ide@vassar.org

Abstract. This tutorial considers Semantic Web representations of linguistically-annotated corpora and related resources — in particular, ontological data — specifically from the perspective of interoperability.

Keywords: Linguistic linked data · Ontologies · Interoperability

1 Outline of the Tutorial

Recent years have seen increasing interest in representing linguistically-annotated data on the Semantic Web in order to link together different resources, including annotated corpora, lexicons, wordnets and framenets, etc., potentially in multiple languages. Queries against such a massively linked data structure could open up possibilities for language and linguistic research that have previously not been feasible. However, despite the development of international standards for the Semantic Web, including W3C standards such as RDF and OWL, creation of linked data representations for linguistic data remains an open challenge. While W3C standards provide *syntactic interoperability* among linked data representations by providing a common physical format, they say nothing about *semantic interoperability* [1], that is, a common set of concepts and relations that can provide a coherent reference for language resources. The existence of a repository or, more usefully, an ontology of linguistic concepts to which different language resources — including annotated corpora, lexicons, and the like — can be linked will be required in order to query these resources and, in general, exploit their interconnections.

This tutorial considers Semantic Web representations — in particular, ontological representations — of linguistically-annotated corpora and related resources from the perspective of rendering them interoperable. The goal is to demonstrate the difficulties of achieving semantic interoperability for linguistic concepts. To that end, students in small groups of 2–4 design an ontological representation for some basic linguistic annotation categories, such as token, part-of-speech, named entities, noun chunks, and the like. To do so it is necessary to determine the relevant objects, their features (properties), and relationships among the objects.[1] The sometimes vast differences

---

[1] Note that consistency among the *names* used for various concepts and relations is not necessary if they describe common concepts, since alternative names for the same object can be trivially mapped.

among the resulting ontologies demonstrates to the students how different the conceptual view of the defined space can be, even among those with very similar backgrounds and perspectives, which in turn highlights the difficulties of establishing a common set of objects and relations that can be agreed upon—and useful to—the creators of different language resources. The differences can be as simple as location in a taxonomy (e.g., Annotation → Syntactic annotation → PhraseStructure vs. Annotation → PhraseStructure), or differences in object-property relations (e.g., Token and Part-of-Speech as objects joined with a "has-a" property, vs. Token with a "part-of-speech" property linked to a category such as noun, verb, etc.); other variations may carve up the conceptual space so differently that comparison is itself problematic.

At the end of the exercise students discuss potential approaches to solving the problem of semantic interoperability for linguistic resources and consider the issues involved, including identification of meaningful vs. trivial differences, means to accommodate different views of the conceptual space, etc.

2   Short Biography

Nancy Ide is Professor and Chair of Computer Science at Vassar College in Poughkeepsie, New York, USA. She has been in the field of computational linguistics for over 30 years and made significant contributions to research in word sense disambiguation, computational lexicography, discourse analysis, and the use of semantic web technologies for language data. She is founder of the Text Encoding Initiative (TEI), the first major standard for representing electronic language data, and later developed the XML Corpus Encoding Standard (XCES) and, most recently, the ISO LAF/GrAF representation format for linguistically annotated data. She has also developed major corpus resources for American English, including the Open American National Corpus (OANC) and the Manually Annotated Sub-Corpus (MASC), and has been a pioneer in efforts toward open data and resources. Professor Ide is Co-Editor-in-Chief of the journal *Language Resources and Evaluation* and Editor of the Springer book series *Text, Speech, and Language Technology*.

Reference

1. Ide, N., Pustejovsky, J.: What does interoperability mean, anyway? toward an operational definition of interoperability. In: Proceedings of the Second International Conference on Global Interoperability for Language Resources (ICGL 2010), Hong Kong (2010)

The Alveo Virtual Lab: Working with an API for Linguistic Data (Invited Tutorial)

Steve Cassidy

Department of Computing, Macquarie University, Sydney, Australia
Steve.Cassidy@mq.edu.au

Abstract. Alveo is a new Virtual Laboratory to support research on Human Communication Science. It provides a repository for language and communication data and an API that provides an interface for tools to work on the data. Some of the goals of the project are to provide a home for data sharing, to make new tools available to researchers in a range of disciplines and to support reproducible research by capturing repeatable workflows for data analysis. This tutorial will describe the main features of the platform and give students the opportunity to work with the API and explore the data holdings in the repository.

Keywords: Corpora · Repository · Language processing · Workflow

1 Outline of the Tutorial

Alveo is a new Virtual Laboratory to support research on Human Communication Science. It is a repository that currently holds more than 15 data sets ranging from small collections of transcribed conversations to a 1000 speaker audio-visual collection of Australian English recordings; new data sets will be contributed by researchers using the platform. Alveo stores the source data (audio, video, text) as well as descriptive metadata and annotations. Since many language resources contain sensitive data and can't be shared on the open web, Alveo requires authentication to access data and offers a level of access control for data owners, allowing them to share data narrowly or widely as appropriate.

An important part of Alveo is a web based (HTTP) API to support query and download of data as well as upload of new annotations. The API follows the principles of Linked Data meaning that every resource has a unique URL and this URL can return a machine readable description of the resource. All data accessed via the API is still subject to authentication and authorization. A variety of language processing and analysis tools have been linked to the API and various libraries and interfaces have been built to make these tools easier to use for non-technical researchers.

One important platform for tool development is the Galaxy Workflow Engine. This is a platform originally developed to support research in biotechnology that has been applied here to allow researchers to build pipelines of language processing tools. Galaxy supports chaining tools together to produce workflows that can be run repeatedly over different data sets. Workflows can then be shared and published to allow others to repeat the same experiment on the same or different data.

This tutorial will give a technical view of the Alveo platform with some detail of the internal models used and a discussion of using the API from Python and R to interface language-processing tools. We will look at running the Galaxy Workflow Engine and adding new tools to that platform. The workshop session will allow students to try out some practical examples and work with audio, video or textual data from Alveo.

2   Short Biography

Steve Cassidy is an Associate Professor in the Computing Department at Macquarie University, Sydney, affiliated with the Centre for Language Technology. His first degree was in Astrophysics, followed by a Masters in Artificial Intelligence from Edinburgh and a PhD in Cognitive Science at Victoria University of Wellington, NZ. While working at the Speech Hearing and Language Research Centre he helped to develop the Emu speech database system and started an interest in corpus based research and managing language resources. Since that time he has worked on standards for Linguistic Annotation, with a particular interest in annotations on multimodal data. He has developed annotation models based on the Resource Description Framework (RDF) and worked on systems to store large-scale collections of language data and annotations in a web-accessible system using Semantic Web technology. Most recently he has been involved in the development of the Alveo Virtual Laboratory that provides a repository for language resources for Australian researchers and an API that interfaces to a range of tools for analysis of language data.

References

1. Cassidy, S., Estival, D., Jones, T., Sefton, P., Burnham, D., Burghold, J.: The Alveo Virtual Laboratory: a web based repository API. In: Proceedings of LREC 2014 (2014)
2. Estival, D., Cassidy, S., Verspoor, K., MacKinlay, A., Burnham, D.: Integrating UIMA with Alveo, a human communication science virtual laboratory. In: Proceedings of OAIF4HLT 2014, Dublin, Ireland (2014)
3. Goecks, J., Nekrutenko, A., Taylor, J.: Galaxy: a comprehensive approach for supporting accessible, reproducible, and transparent computational research in the life sciences. Genome Biol. 11(8), R86 (2010)

Multilingual Linked Data Generation from Language Resources (Invited Tutorial)

Asunción Gómez-Pérez, Jorge Gracia, and Daniel Vila-Suero

Ontology Engineering Group (OEG), Universidad Politécnica de Madrid, Spain
{asun, jgracia, dvila}@fi.upm.es

Abstract. This document briefly describes the tutorial on multilingual linked data generation from language resources imparted at EUROLAN'15. In such course, a theoretical overview of linguistic linked data was given, followed by a hands-on session that covered the main steps and aspects in the lifecycle of linguistic linked data generation and publication. The focus was very practical and the participants had the opportunity of generating linked data of some sample resources by themselves.

Keywords: Linguistic linked data · Lemon · Language resources

1 Outline of the Tutorial

This document briefly describes the invited tutorial imparted by the authors at the 12th Eurolan Summer School on July 2015. The tutorial focused on the process of generating and publishing linked data (LD) from a domain specific multilingual resource. The main activities of the LD generation process [1] were presented, as well as some tools for multilingual LD generation and linking. A technical description of these and other related techniques can be found in [2].

Prior to the practical session, a theoretical introduction was given in order to motivate the problem and introduce the basic concepts used later during the hands-on session. In such a talk, the lecturers reviewed some typical use cases for language resources discovery and pointed out the limitations of current techniques. Then, LD was presented as a solution for these and other problems related to linguistic data integration. Also, the current efforts of some European projects (e.g., Lider[1]) and W3C community groups (BPMLOD, LD4LT and Ontolex) in achieving the vision of a multilingual Web of Data were described.

During the hands-on session that followed, participants were guided in the particular task of completing the transformation of a multilingual resource (e.g., a bilingual dictionary) into LD[2]. The session started by illustrating the methodology to follow with a real example, based on the Apertium RDF use case[3]. Then, the lecturers provided a

---

[1] http://www.lider-project.eu/

[2] Following the guidelines at http://bpmlod.github.io/report/bilingual-dictionaries/index.html

[3] http://linguistic.linkeddata.es/apertium/

series of simple steps and best practices for: (1) the analysis, cleaning and transformation of the source data into RDF by using generic tools such as Open Refine[4], mapping it to existing vocabularies such as lemon [3], (2) discovering links to other LD resources using a reconciliation service, (3) publishing the produced RDF data on the Web by using a LD front-end and a SPARQL endpoint, and (4) using SPARQL to query the data that has been made available.

2 Short Biographies

Prof. Dr. Asunción Gómez-Pérez is Full Professor at Univ. Politécnica de Madrid (UPM) where she directs the OEG group. She participated in 21 EU projects being the coordinator of some of them (OntoGrid, SemSorGrid4Env, SEALS, Lider). She has published more than 150 papers and co-authored two books in the field of Ontological Engineering. She has been co-director of the summer school on Ontological Engineering and the Semantic Web since 2003. She has been program chair and organizer of many relevant conferences and workshops on ontologies and Semantic Web.

Dr. Jorge Gracia is a postdoctoral researcher at the Artificial Intelligence Department at Univ. Politécnica de Madrid (UPM). He has worked in several EU projects: Dynalearn, Monnet, and LIDER. His main research interests are ontology matching, query interpretation, and multilingualism on the Web of Data. He has co-organised several tutorials, workshops, and a datathon in the field of linguistic LD.

Mr. Daniel Vila-Suero is a PhD student at the OEG, and MSc in Computer Science. His research topics are multilingualism in the Web of Data, methodologies, digital libraries and LD. He participated in several Spanish research projects related to LD and multilingualism. He is currently participating in the European project Lider.

References

1. Vila-Suero, D., Gómez-Pérez, A., Montiel-Ponsoda, E., Gracia, J., Aguado-de Cea, G.: Publishing linked data: the multilingual dimension. In: Buitelaar, P., Cimiano, P. (eds.) Towards the Multilingual Semantic Web, pp. 101–118. Springer, Berlin (2014)
2. Gracia, J., Vila-Suero, D., McCrae, J., Flati, T., Baron, C., Dojchinovski, M.: Language resources and linked data: a practical perspective. In: Lambrix, P., Hyvönen, E., Blomqvist , E., Presutti, V., Qi, G., Sattler, U., Ding, Y., Ghidini, C. (eds.) EKAW 2014. LNCS, vol. 8982, pp. 3–17. Springer International Publishing, Switzerland (2014)
3. McCrae, J., Aguado-de Cea, G., Buitelaar, P., Cimiano, P., Declerck, T., Gómez-Pérez, A., Gracia, J., Hollink, L., Montiel-Ponsoda, E., Spohr, D., Wunner, T.: Interchanging lexical resources on the semantic web. Lang. Resour. Eval. **46**, 701–719 (2012)

---

[4] http://openrefine.org/

Annotation Interoperability
(Invited Tutorial)

Christian Chiarcos

Goethe Universität Frankfurt, Germany
chiarcos@informatik.uni-frankfurt.de

Abstract. The tutorial addressed the problem of heterogeneous representations for language resources, in particular annotated corpora. Heterogeneity exists on multiple levels, the most important being the levels of physical representation, and annotation vocabulary. Establishing interoperability between and among corpora, dictionaries and NLP tools thus involves two dimensions, as well, namely *structural interoperability* (defining how to access the data, i.e., common formats and protocols) and *conceptual interoperability* (defining how to interpret the data, i.e., by reference to a common vocabulary). In this tutorial, both aspects were addressed with a focus on linguistic annotations and the *Ontologies of Linguistic Annotation* (OLiA). As a means to solve the problem of conceptual interoperability for annotations, the OLiA ontologies represent a major hub of linguistic terminology in the Linguistic Linked Open Data (LLOD) cloud. Finally, use cases from natural language processing and corpus querying were described.

Keywords: Linguistic annotations · Ontologies · OLiA (Ontologies of Linguistic Annotation) · Corpus linguistics · Natural language processing

1 Outline of the Tutorial

Limited interoperability between annotations across different annotation schemes both within as well as across languages represents a major hurdle in the development of truly interoperable NLP tools and interoperable corpus querying. By leveraging multiple terminology repositories maintained by different communities (e.g., ISOcat[1], GOLD[2], MULTEXT/East[3], Typological Database System[4]) as well as annotation schemes applied to more than 70 languages, the Ontologies of Linguistic Annotation (OLiA, [2]) provide a solution for annotation interoperability for corpora and NLP tools in the context of the Linguistic Linked Open Data (LLOD) cloud.

The OLiA ontologies provide a vocabulary of cross-linguistically applicable terms and concepts as used in the linguistic annotation for various linguistic phenomena (primarily morphosyntax, syntax, discourse, coreference and information structure, as

---

[1] http://www.isocat.org/rest/dc/
[2] http://linguistics-ontology.org
[3] http://nl.ijs.si/ME/owl/
[4] http://languagelink.let.uu.nl/tds

found in publicly available annotated corpora) on a great band-width of languages. The OLiA ontologies are available under an open license from http://purl.org/olia. A development version with experimental additions and tools is available from http://sourceforge.net/projects/olia/.

The Ontologies of Linguistic Annotations represent a modular architecture of OWL2/DL ontologies that formalize the mapping between annotations, a 'Reference Model' and existing terminology repositories ('External Reference Models'). In the OLiA architecture, four different types of ontologies are distinguished:

- The OLIA REFERENCE MODEL specifies the common terminology that different annotation schemes can refer to. It is derived from existing repositories of annotation terminology and extended in accordance with the annotation schemes that it was applied to.
- Multiple OLIA ANNOTATION MODELs formalize annotation schemes and tagsets. Annotation Models are based on the original documentation, so that they provide an interpretation-independent representation of the annotation scheme.
- For every Annotation Model, a LINKING MODEL defines rdfs:subClassOf relationships between concepts/properties in the respective Annotation Model and the Reference Model. Linking Models are interpretations of Annotation Model concepts and properties in terms of the Reference Model.
- Existing terminology repositories with an RDF/OWL representation (e.g., ISOcat and GOLD) are integrated in this architecture as EXTERNAL REFERENCE MODELs. Linking Models specify the relationships between Reference Model and External Reference Models analogously to the relation between Annotation Model and OLiA Reference Model.

Being developed over more a decade [10], OLiA has been used to facilitate interoperability and information integration of linguistic annotations in corpora, NLP pipelines, and lexical-semantic resources in various applications. In addition, the tutorial touched innovative, ontology-based approaches for NLP tasks such as cross-tagset morphosyntactic analysis.

The tutorial featured four exemplary use cases:

- documenting and re-analyzing tagsets and meta-tagsets [6]
- concept-based querying over linguistic corpora [5]
- modeling and querying multi-layer corpora with RDF and SPARQL [4]
- ontology-based ensemble combination [3]

Additional use cases not covered in the tutorial include interoperable specifications for NLP pipelines [1, 9], annotation projection [7], ontology-based morphosyntactic analysis with neural networks [11], and uses of OLiA in information extraction [8].

In the accompanying hands-on session, participants implemented new OLiA Annotation Models and their linking to the OLiA Reference Model. In the session, we converted the language-independent specifications of the Universal Dependencies (UD)[5] to OWL2/DL (to be used as yet another external reference model), we produced

---

[5] http://universaldependencies.github.io/docs/

Annotation Models for language-specific versions of the Universal Dependencies and we linked these with the OLiA Reference Model. Discussions to integrate these models in the UD infrastructur are currently being conducted.

2 Short Biography

Christian Chiarcos is Assistant Professor of Computer Science at Goethe University Frankfurt, Germany, and heading the Applied Computational Linguistics group. In 2010, he received a doctoral degree on Natural Language Generation from the University Potsdam, Germany. He subsequently worked at the Information Sciences Institute of the University of Southern California (ISI/USC), before joining Goethe University in 2013. His research focuses on semantic technologies, including computational semantics as well as Semantic Web technologies. Specific interests cover computational discourse semantics (machine reading), NLP and language resource interoperability. As a computational linguist, Christian Chiarcos explored Semantic Web and Linked Data from an NLP perspective and contributed to the emergence of a community at the intersection of both areas: He has been co-founder of the Open Linguistics Working Group of the Open Knowledge Foundation (OWLG), he initiated and co-organized both the Linked Data in Linguistics workshop series and the accompanying development of the Linguistic Linked Open Data cloud.

References

1. Buyko, E., Chiarcos, C., Pareja Lora, A.: Ontology-based interface specifications for an NLP pipeline architecture. In: Proceedings of the 6th Language Resource and Evaluation Conference (LREC-2008), Marrakech, Morocco, May 2008, pp. 847–854 (2008)
2. Chiarcos, C.: An ontology of linguistic annotations. LDV Forum (GLDV-J. Comput. Linguist. Lang. Technol.) **23**, 1–16 (2008)
3. Chiarcos, C.: Towards robust multi-tool tagging. An OWL/DL-based approach. In: Proceedings of the 48th Annual Meeting of the Association for Computational Linguistics. Uppsala, Sweden, July 2010, pp. 659–670 (2010)
4. Chiarcos, C.: Interoperability of corpora and annotations. In: Chiarcos, C., Nordhoff, S., Hellmann, S. (eds.) Linked Data in Linguistics. Representing and Connecting Language Data and Language Metadata. Springer, Heidelberg (2012)
5. Chiarcos, C., Dipper, S., Götze, M., Leser, U., Lüdeling, A., Ritz, J., Stede, M.: A flexible framework for integrating annotations from different tools and tagsets. TAL (Traitement automatique des langues) **49** (2008)
6. Chiarcos, C., Erjavec, T.: OWL/DL formalization of the MULTEXT-east morphosyntactic specifications. In: Proceedings of the 5th Linguistic Annotation Workshop (LAW-V), held in conjunction with the ACL-HLT 2011, June 2011. Portland, Oregon, pp. 11–20 (2011)
7. Chiarcos, C., Sukhareva, M., Mittmann, R., Price, T., Detmold, G., Chobotsky, J.: New technologies for old germanic. Resources and research on parallel gospels in older continental western germanic. In: 8th Workshop on Language Technology for Cultural Heritage, Social Sciences, and Humanities (LaTeCH-2014), held in conjunction with EACL-2014. Gothenburg, Sweden, April 2014, pp. 22–31 (2014)

8. Hahm, Y., Lim, K., Park, J., Yoon, Y., Choi, K.-S.: Korean NLP2RDF resources. In: Proceedings of the 10th Workshop on Asian Language Resources, held in conjunction with COLING-2015, Dec., Mumbai, India, pp. 1–10 (2012)
9. Hellmann, S., Lehmann, J., Auer, S., Brümmer, M.: Integrating NLP using linked data. In: Alani, H., et al. (eds.) The Semantic Web–ISWC 2013. LNCS, vol. 8219, pp. 98–113. Springer, Berlin (2013)
10. Schmidt, T., Chiarcos, C., Rehm, G., Lehmberg, T., Witt, A., Hinrichs, E.: Avoiding data graveyards: from heterogeneous data collected in multiple research projects to sustainable linguistic resources. In: Tools and Standards - The State of the Art. Proceedings of the E-MELD 2006 Workshop on Digital Language Documentation. Michigan State University in East Lansing, Michigan (2006)
11. Sukhareva, M., Chiarcos, C.: An ontology-based approach to automatic part-of-speech tagging using heterogeneously annotated corpora. In: Proceedings of the Second Workshop on Natural Language Processing and Linked Open Data (NLP&LOD2), collocated with RANLP 2015, Hissar, Bulgaria, pp.23–32 (2015)

Linked Data for Lexicons
(Invited Tutorial)

John P. McCrae

Cognitive Interaction Technology, Center of Excellence Bielefeld University
and Insight Centre for Data Analytics, National University of Ireland Galway
john@mccr.ae

Abstract. Lexicons form a crucial part of how we build natural language systems that allow humans to interact with machines and to build web applications that can use web standards such as OWL but express them in natural language, we developed a vocabulary called *lemon* (Lexicon Model for Ontologies). This tutorial details the model and enables participants to apply it in line with common patterns of usage.

Keywords: Lexicons · Semantic Web · Linked Data · Ontologies

1 Outline of the Tutorial

Lexicons, machine-readable dictionaries, wordnets and similar resources are an important tool for a wide range of applications in natural language processing and it is important that they are linked to both corpora and semantic resources such as ontologies. For this reason, the OntoLex/*lemon* (Lexicon Model for Ontologies) model [1] was developed to enable the representation of entries in a lexicon and the representation of their forms, part-of-speech and other syntactic information in combination with semantic information found in ontologies on the Web. Thus, this model gives meaning to words in a formal language understandable to machines as well as grounding them in instances in corpus.

This lecture covers the basics of the *lemon* model, in particular the core module, as well as the modules covering representation of the syntax-semantics correspondences by means of frames, the use of decompositions for representing multiword terms, the inclusion of arbitrary properties and relationships in the model and finally the representation of metadata by means of the Lime Module [2]. We then covered the LexInfo model [3], which includes extended properties and classes to further enable the representation of lexical information.

The second part of the lecture covers the common patterns as expressed by the *lemon* Design Patterns Language [4], representing the major usage of the model. For nouns, we covered the lexicalization of named entities, common nouns that reference classes and nouns that express relationships. For verbs, we covered both static verbs that express factual relationships and event verbs that can be used to describe complex ontology frames with many properties. Finally, for adjectives, we focused primarily on intersective adjectives, but also briefly saw the modeling of scalar adjectives. The formal syntax of the design pattern language was introduced and this was used as the basis of a hands-on activity, where participants created their own *lemon* lexicons.

Finally, the adoption of *lemon* was demonstrated, in particular the BabelNet model [5], which is the focus of another talk presented at this EUROLAN summer school.

2   Short Biography

John P. McCrae completed an MSc in Mathematics and Computer Science from Imperial College London, followed by a PhD at the National Institute of Informatics, Tokyo. Since 2009, he worked at the Cognitive Interaction Technology Excellence Center at University Bielefeld, where he has been involved in projects on multilingual ontologies, portable dialogue systems and multilingual linked data. In that time, he has worked to create a growing cloud of linguistic linked data by means of developing the *lemon* (Lexicon Model for Ontology) model and organizing workshop series including the Linked Data in Linguistics, Multilingual Linked Open Data for Enterprise and Multilingual Semantic Web workshops. In August 2015 he started a new position as a research fellow at the Insight Centre for Data Analytics at the National University of Ireland Galway.

References

1. McCrae, J., Aguado-de-Cea, G., Buitelaar, P., Cimiano, P., Declerck, T., Gómez-Pérez, A., Wunner, T.: Interchanging lexical resources on the Semantic Web. Lang. Resour. Eval. **46**(4), 701–719 (2012)
2. Cimiano, P., Buitelaar, P., McCrae, J., Sintek, M.: LexInfo: a declarative model for the lexicon-ontology interface. Web Seman. Sci. Serv. Agents World Wide Web **9**(1), 29–51 (2011)
3. Fiorelli, M., Stellato, A., McCrae, J.P., Cimiano, P., Pazienza, M.T.: LIME: the metadata module for OntoLex. In: Gandon, F., Sabou, M., Sack, H., d'Amato, C., Cudré-Mauroux, P., Zimmermann, A., (eds.) The Semantic Web. Latest Advances and New Domains, pp. 321–336. Springer International Publishing, Switzerland (2015)
4. McCrae, J.P., Unger, C.: Design patterns for engineering the ontology-lexicon interface. In: Buitelaar, P., Cimiano, P. (eds.) Towards the Multilingual Semantic Web, pp. 15–30. Springer, Berlin (2014)
5. Ehrmann, M., Cecconi, F., Vannella, D., Mccrae, J.P., Cimiano, P., Navigli, R.: Representing multilingual data as linked data: the case of BabelNet 2.0. In: Proceedings of the 9th International Conference on Language Resources and Evaluation, pp. 401–408 (2014)

Building Ontolex Resources Using Protégé (Invited Tutorial)

Thierry Declerck

Austrian Center for Digital Humanities, Austria and DFKI GmbH,
Multilingual Technologies Lab, Germany
declerck@dfki.de

Abstract. In this tutorial the aim was to introduce both to the use of the Protégé ontology editor and to the building of lexical resources on the basis of an ontological model for such resources, the Ontolex model, which has been developed in the context of a W3C Community Group. The overall goal was to show how lexical data can be encoded in such a way that they can be published in the emerging Linguistic Linked (Open) Data.

Keywords: Linguistic Linked Open Data · Ontolex · Protégé

1 Outline of the Tutorial

In this tutorial, the aim was to introduce EUROLAN students to the building of lexical resources on the basis of an ontological model for such resources, the Ontolex model, which has been developed in the context of a W3C Community Group[1]. We introduced also to the Protégé ontology editor[2], which was used for the manual encoding of lexical resources in the Ontolex format. The overall goal was to show how lexical data can be encoded in such a way that they can be published in the emerging Linguistic Linked (Open) Data[3] framework.

In a first step, we introduced some main concepts of what constitutes an ontology in the Semantic Web context. We described therefore elements like classes and the subclass hierarchy, together with the concept of *instance* (or *individual*) of a class. We further introduced the concept of *property*, which can be either an object property (relating two objects of the ontology) or a data-type property, adding a literal value to an ontology object. Stress was put on the fact that all objects of the ontology are associated with an URI, making them thus linkable in the Linked (Open) Data context. We also introduced to the usage of the visualization plug-ins for Protégé, especially OntoGraf[4], in order to support an easier understanding of the Ontolex model.

---

[1] https://www.w3.org/community/ontolex/
[2] http://protege.stanford.edu/
[3] http://linguistic-lod.org/llod-cloud
[4] http://protegewiki.stanford.edu/wiki/OntoGraf

2 Short Biography

Thierry Declerck is senior consultant at the Language Technology Lab of DFKI GmbH (www.dfki.de/lt), being working for DFKI since June 1996. He is currently also member of the Department of Computational Linguistics Lab at Saarland University, where he is leading the University's contribution to the European FP7 project PHEME (http://www.pheme.eu/). Thierry Declerck joined the DFKI's Language Technology Lab in June 1996. He is now contributing to the LIDER (http://www.lider-project.eu/), dealing with the establishment of the Linguistic Linked Open Data (LLOD) cloud, and to FREME projects (http://www.freme-project.eu/), dealing with application scenarios for the LLOD. Before this, he was the coordinator of the European FP7 project TrendMiner (http://www.trendminer-project.eu/), which very successfully terminated in October 2014. Thierry Declerck was an active member of ISO TC37/SC4 and DIN NAAT6. He was, for example, editor of the ISO standard for syntactic annotation (SynAF), which has been submitted as a final draft for international standard in June 2010. More recently, he is involved in W3C community groups, like OntoLex. Since February 2010, Thierry Declerck is occasional consultant for Austrian Academy of Sciences, mainly for ensuring the tight connection between language technology and semantic web technologies in the eHumanities. Thierry Declerck has also been involved in many program committees and co-chairing workshops and conferences. He is for example co-chair for the LREC conferences, since 2012.

Acknowledgments. This tutorial was supported in part by the European Union, both by the LIDER project (under Grant No. 610782), and by the FREME project (under Grant No. 644771).

This work was conducted using the Protégé resource, which is supported by grant GM10331601 from the National Institute of General Medical Sciences of the United States National Institutes of Health.

Reference

1. Chiarcos, C., Cimiano, P., Declerck, T., McCrae, J.P.: Linguistic linked open data (LLOD) - introduction and overview. In: Chiarcos, C., Cimiano, P., Declerck, T., McCrae, J.P. (eds.) 2nd Workshop on Linked Data in Linguistics, Pisa, Italy, CEURS, pp. i–xi (2013)

Needles in a Haystack and How to Find Them. Building a Resource to Help Authors (Speakers/Writers) to Overcome the Tip of the Tongue Problem (Invited Tutorial)

Michael Zock

University Aix-Marseille, France
Michael.Zock@lif.univ-mrs.fr

Abstract. Whenever we speak, read or write we always use words, the exchange money of concepts they are standing for. No doubt, words ARE important. Yet having stored 'words' does not guarantee that we can access them under all circumstances. Some forms may refuse to come to our mind when we need them most, the moment of speaking or writing. This is when we tend to reach for a dictionary, hoping to find the token we are looking for. The problem is that most dictionaries, be they in paper or electronic form, are not well suited to support the language producer.

Hence the questions, why is this so and what does it take to enhance existing resources? Can we draw on what is known about the human brain or its externalized form (texts)? Put differently, what kind of help can we expect by looking at the work done by neuroscientists, psycholinguists or computational linguistics? These are some of the questions I will briefly touch upon, by ending with a concrete proposal (roadmap), outlining the majors steps to be performed by drawing heavily on corpus linguistics to build and combine resources in order to enhance an existing electronic dictionary.

Keywords: Word access · Mental lexicon · Associative network · Index

1   Outline of the Tutorial

Everyone has encountered already the following problem: we look for a word or the name of someone (friend, actor, …) or something (tree, capital, …), without being able to find the corresponding form (name, lemma, i.e. word form).[1] Psychologists have studied this problem extensively with respect to word access. This work has shown that people being in the Tip-of-the-tongue- state know many things concerning the elusive word (meaning, number of syllables, origin, etc.), and the words coming to their mind,

---

[1] This is somehow similar to the problem we encounter when searching for our keys, spectacles, or any object you can think of. In all these cases, we tend to know something concerning the eluding object: shape, color, smell, time (when we've used it) space (where we've met or where we've put it this morning), or related objects, etc.

words they tend to confuse the target with, look strangely similar to the target in terms of meaning (left vs. right) or form (syntactic category, phonemes/syllables).

Word access difficulties are similar to other search problems. While there are obvious differences between words, keys, and spectacles, when it comes to *search* and *organization* there are many similarities between them. To find these 'objects' they must exist, we must know where they are stored, i.e. their location must be known (access key, index) and the organization must be somehow rational.[2] Obviously, there are also some differences between objects and words. For example, when searching an object we can usually name the target. This is not so in the case of word-finding problems, as if we could name the target, there would be no search problem. This being said, in both cases we do need an index in particular if the number of items grows. Think of a huge department store devoid of this help. The customer would be completely lost if he were not given a category list telling him roughly where to find what (food, cosmetics, sportswear, …).

Language users have to some extent this kind of metaknowledge. They generally know the category to which the target belongs, they do know its meaning and many other 'details'. Since authors know so many things concerning the target, I suggest to take advantage of this fact and start from there. Put differently, my goal is to create a resource allowing a user to start from his knowledge, no matter what it may be,[3] leading him from this input to the desired output, the elusive word. To this end, I plan to add to an existing electronic dictionary an index based on associations (collocations encountered in a large corpus). In other words, I propose to build a dictionary functionally similar to the mental lexicon. This being said, in our resource words are holistic entities, unlike the words in the human brain, they are not decomposed. The function of the words provided by the user (input) is to convey meanings or meaning fragments, i.e. conceptually or collocationally related words. Hence our resource is meant to contain next to conventional information (meaning) links to related words, allowing access of the target via meanings or meaning fragments or somehow related, cooccuring terms (black → coffee). The approach I am proposing is based on the following assumptions.

1. Search strategies depend on the way how words are represented in our mind. Even though we still lack a precise map of the mental lexicon (the way how words are represented and organized in our mind), we do know enough about it to build an analogue resource. Concerning search there are two main strategies, operating on two orthogonal axes.

- The first one goes from *meaning to forms*. This is the natural order of 'things': starting from some underspecified conceptual input (meaning, mental image) we go towards its expression (sound or graphic form of the word) via some intermediate terms (lemma or lexical concepts in the theory of Levelt).

---

[2] For example, similar objects should be put in the same area. Similarity can, of course, be a matter of viewpoint. Note, that while difficult in real world, in an abstract representation an object can occur in many places (functional view, taxonomic view, etc.).

[3] Note that knowledge may vary from person to person and from moment to moment.

- The second axis is more sensitive to what can we can call the organization of the mental lexicon. Words are linked in terms of usage, that is, words typically occurring together are encoded as neighbors in the network (co-occurrence network, i.e. network of associations). Hence to find the target, we consider not only intrinsic information concerning this specific word (typically elements of the definition), but also cooccurrent words (strong → coffee).

In sum, we have two complementary approaches: one where we start from *meaning* (idea, concept), **plan-A**, the other where we start from a somehow related word (cooccurrence), **plan-B**. This latter may be related to the meaning, i.e. be part of the definition or the target form, in any case our goal. Put differently, to find the elusive word we consider not only its intrinsic meaning or meaning fragments, but also related words (neighbors, co-actors), i.e. other words playing a role in the scenario (sentence) in which the target occurs (*target* : mocha; related words : coffee, black, ...).

If the first axis is the golden route, the natural way in which production occurs (Plan A), the second axis (associative way) is an alternative (Plan-B), route used in case of failure. The first process is automatic, fast and unconscious, while the second is deliberate, slow and accessible to our awareness. This is the one I am interested in, because this is the one an author is confronted with when reaching for an external resource (dictionary or thesaurus).

2. The representation of the mental lexicon (or its equivalent) is a vast network whose nodes are concepts or words (lemmas, idioms) and whose links are essentially associations. Since everything is connected, everything can be accessed (from any-where), at least in principle. All we need is an input and then follow the (right) links. This is tantamount of saying that searching a word consists mainly in entering the network by providing a term (knowledge available at that moment, generally a term more or less directly associated to the target) and follow then the links until one stumbles upon the elusive word. If the source word (prime) is a direct neighbor of the target, we will immediately find the desired word in a (flat) list, otherwise we have to repeat the operation by using a different query term (new source word).

3. The mental dictionary is both a *dictionary* and an *encyclopedia*. Actually, this distinction does not really make sense anymore. Since words are used to encode *knowledge of the world*, this latter can help us to find the eluding word. For example, the term 'chopstick' could well be accessed via 'Chinese restaurant' or via 'oriental tableware'. Everything being associated with something, everything can be evoked by something. Of course, not everything is a direct neighbor. Hence there may be occa-sions where we have to perform search in several steps.

4. The information needed to allow for this type of navigation (semantic map) can be found both in the brain (associations) as well as in its externalized form, language products (speech errors, corpora). Being an externalization of the way how our thoughts (concepts/words) are organized in our brain, they are a valuable clues con-cerning the human mind. This should allow us to create a similar model, a kind of atlas or semantic map allowing users to navigate, moving from a starting point (knowledge available at a given moment, word) to the goal (target word).

Here is in a nutshell the method, i.e. roadmap for achieving this. Again, my ultimate goal is to help authors (speakers, writers) to overcome the tip-of-the-tongue-problem.

Whenever we need a word, we look it up in the place where it is stored, the dictionary or the *mental lexicon*. Since storage does not guarantee access we must develop a strategy allowing us nevertheless to locate the elusive word. To do so we tap on the user's knowledge, (ideally) whatever it may be, (in our case) mainly concepts or words associated with the target. In sum, we suggest to model lexical access as a kind of dialog between a user and the system (lexical data-base).

The process consists basically in the following steps : (a) the user iniates search via some input (query), (b) the system answers by presenting a list of potential candidates (output) and (c) the user decides then whether any of these candidates matches his goal, or if not, with what specific word to continue search with. Concretely speaking this could look like this. The user starts by providing her input, that is, any word coming to her mind, word somehow connected to the target (step-1). The system presents then in a clustered and labeled form (categorial tree) all direct associates (step-2).[4] The user navigates in this tree, deciding on the category within which to look for the target, and if he cannot find it in any of them, in what direction to go. If he finds the target, search stops, otherwise he will pick one of the associated terms or provide an entirely new word and the whole process iterates. The system will come up with a new set of proposals.

2   Short Biography

Michael Zock, born in Germany, has been living in France ever since 1970. He gained there a Ph.D in experimental psychology (psycholinguistics). Since 1989 he has been a researcher at the CNRS, working in the Language & Cognition group of LIMSI, an AI laboratory close to Paris (Orsay). Currently he is emeritus research director at the Laboratoire d'Informatique Fondamentale (LIF) at Marseille. His research interests lie in the modeling of the cognitive processes underlying natural language generation and in the building of tools for language learning (CALL systems). The focus of his recent research has been on outline planning (help authors to perceive possible links between their ideas in order to produce coherent discourse), lexical access (help authors to find an elusive word by taking into account certain features of the mental lexicon), and the acquisition of basic speaking skills: help students to become quickly fluent in a foreign language by learning the basic vocabulary and syntactic structures (learn words in context).

---

[4] This labeling is obligatory to allow for realistic navigation, as the list produced in responsive to the input may be very long and the words being of the same kind may be far apart from each other in the list. Hence it makes sense to structure words into groups by giving them appropriate (i.e. understandable) names so that the user, rather than looking up the entire list of words, searches only within a specific bag labeled by a category.

Towards Interoperability of Annotation. Use Cases of Corpora and Software (Invited Tutorial and Practical Session)

Dan Cristea and Andrei Liviu Scutelnicu

Alexandru Ioan Cuza University of Iaşi, Faculty of Computer Science Romanian Academy, Institute for Computer Science
{dcristea, liviu.scutelnicu}@info.uaic.ro

Abstract. The tutorial and practical session is dedicated to interoperability of text annotation at two levels: discourse and lexicographic resources. The practicalities of these types of annotations are related both to theoretical achievements (improving the representation of discourse) and to practical ones (improving the behaviour of discourse parsers, of semantic text analytics programs, and of programs relating terms and text mentions for the purpose of developing better search results and acquiring related information in external sources).

Keywords: Discourse representation · Discourse parsing · Discourse annotation · Annotated corpora · Lexicographic resources · Interoperability of resources

1 Outline of the Tutorial

The tutorial presented, in five tempos, issues related to the interoperability of text annotation and software exploiting it. Part one was an overview over discourse related matters. Three influential discourse theories (Rhetorical Structure Theory, Centering and Veins Theory [1]) were briefly presented, with more emphasis on the last one and its potential power to model the way humans read, memorise and summarise a text. The second part was dedicated to the MASC [2] experiment – a corpus of American English[1], which was recently updated to include discourse markings: discourse unit boundaries, discourse cue-words and nuclearity. The expectation is that the conventions of annotation introduced will prove effective for an underdetermined representation of the discourse structure, more economical to be handled by human annotators and prone to fewer errors in an automatic parsing process. The third part reported on the *QuoVadis* [3] experiment – a corpus based on the Romanian version of the well-known "Quo Vadis" novel, by Henryk Sienkiewicz, to include annotations of entities and semantic relations between them (referential, kinship, affective and social). The corpus will be used to train programs to identify character mentions and to link them, evidencing thus relations able to allow intelligent searches in the content and to

---

[1] See the lecture given by Nancy Ide, Vassar College, abstract in this volume.

visualise content-based statistics. The forth part presented *MappingBooks* [4] – an on-going project at UAIC aimed to link mentions of events/locations/persons of books in the virtual and real world, that allows the reader to get selective and instantaneous information from virtual sites, also in relation with her/his own instantaneous coordinates. Finally, we showed a way to solve an instance of the tip-of-the-tongue problem[2] [5]. By exploiting the inherent linear/hierarchical organisation and their modern XML serialization, lexicographic and other types of linguistic resources can be encapsulated in shells that make them interoperable. We were focused on Romanian resources (a dictionary, a wordnet, a corpus and a lexicon), but the approach is general enough to be extended to other types of resources and languages.

2 Short Biographies

Dan Cristea is a professor at the "Alexandru Ioan Cuza" University of Iaşi (UAIC), Faculty of Computer Science (FII). Years ago he has initiated a line of research in CL and NLP at UAIC-FII, which has grown in time both numerically and qualitatively. He is the initiator of the series of EUROLAN Summer Schools in NLP. He is known for his work on discourse processing, multilingual linguistic workflows, computational lexicography, and Romanian linguistic resources.

Andrei Liviu Scutelnicu has got a degree in Computer Science and a master in Computational Linguistics at UAIC-FII, and is currently developing a PhD thesis on conceptual and technical solutions of linking linguistic resources of different types. Both authors also have part-time positions at the Institute of Computer Science of the Romanian Academy in Iaşi.

References

1. Cristea, D., Ide, N., Romary, L.: Veins theory. a model of global discourse cohesion and coherence. In: Proceedings of Coling 1998 and ACL 1998, Montreal, August, pp. 281–285 (1998)
2. Ide, N., Baker, C., Fellbaum, C., Fillmore, C., Passonneau, R.: MASC: the manually annotated sub-corpus of American English. In: Proceedings of the Sixth Language Resources and Evaluation Conference (LREC), Marrakech, Morocco (2008)
3. Cristea, D., Gîfu, D., Colhon, M., Diac, P., Bibiri, A.-D., Mărănduc, C., Scutelnicu, L.-A.: Quo Vadis: a corpus of entities and relations. In: Gala, N., Rapp, R., Enguix, G.B. (eds.) Language Production, Cognition, and the Lexicon. Springer, Switzerland (2015)
4. Cristea, D., Pistol, I.-C.: MappingBooks: linguistic support for geographical navigation systems. In: Colhon, M., Iftene, A., Barbu-Mititelu, V., Cristea, D., Tufiş, D. (eds.) Proceedings of the 10th ConsILR Conference, Craiova, 18–19 September, pp. 189–198. "Alexandu Ioan Cuza" University Publishing House (2014)
5. Zock, M., Cristea, D.: You shall find the target via its companion words: specifications of a navigational tool to help authors to overcome the tip-of-the-tongue problem. In: Proceedings of 11th International Workshop on Natural Language Processing and Cognitive Science, Venetia, 27–29 October 2014

---

[2] See the lecture given by Michael Zock, abstract in this volume.

Challenges in Building a Publicly Available Reference Corpus (Invited Tutorial)

Dan Tufiş

Research Institute for Artificial Intelligence "Mihai Drăgănescu",
Romanian Academy
tufis@racai.ro

Abstract. One of the most important decisions of a NLP community is building a reference corpus for the language in case. It is a scientifically exciting, multidisciplinary project and it has a major cultural dimension. In an IPR strictly regulated society, gathering large quantities of text and speech data, representative for a language is not an easy task.

The talk took the participants to a journey into the intricacies, challenges and opportunities of building a large IPR-cleared reference corpus, intended for open access. It was a mildly introduction to major decision makings concerning the corpus structure and size, its encoding, the choice of the corpus management platform, types of services to be offered to the corpus users, current and future processing and storage technologies. Although the case study refers to Romanian, the shared experience, expected comments and suggestions will be relevant for initiatives devoted to other languages.

Keywords: Corpus · IPR resolution · Data acquisition and sampling · Language representativity · Metadata · Text and speech annotation · Text-to-speech synthesis

1 Outline of the Tutorial

The tutorial starts with clarifying the distinctions between various collections of linguistic data (archives, electronic text libraries and corpora) and discusses fundamental concepts of representative corpora: representativeness, balance and sampling. Then the major questions to be answered before starting a project on building a large and representative corpus are suggested and commented. The discussion is exemplified by the structures (population and distribution) of Brown Corpus and British National Corpus. The issue of annotation is presented in some details and exemplified with pros and cons for inline, standoff and hybrid solutions.

The second part of the tutorial was dedicated to the presentation of the COROLA (COntemporary ROmanian LAnguage) corpus and answers to the previously raised questions: what is its main purpose and the envisaged users, what are the linguistic levels encoded into the annotations, the metadata schema, the exploitation platform and the services it offers, the tools used for text and speech annotation, the structure of the corpus (domains and styles) and the quantitative distribution of the population as planned for the first version. There were discussed at some length the discussions and the final cooperation agreements concluded with all linguistic data providers (publishing houses, newspapers and journals publishers, national and local radio broadcasters and bloggers). In the final part of the tutorial there were presented the current statistics of the already processed, indexed and stored linguistic data (more than 100 Mwords of textual data and about 17 hours of transcribed and phoneme aligned speech). The processing chains (sentence splitting, tokenization, POS tagging, lemmatization, chunking and dependency linking) were exemplified.

The hands-on session was dedicated to demonstrating the speech processing workflow and the quality text-to-speech production.

2 Short Biography

Dan Tufiş is a member of the Romanian Academy and the director of the Research Institute for Artificial Intelligence "Mihai Drăgănescu" of the Romanian Academy and a Honorary Professor of the University "A.I. Cuza" of Iaşi. He was one of the pioneers of NLP in Romania, working in this area for more than 35 years. He authored more than 250 peered-reviewed papers and has many contributions in the area of computational morphology (paradigmatic morphology), machine readable dictionaries (WebDEX) including wordnets (Ro-wordnet), corpus processing (tiered tagging, word alignment, question answering, sentiment analysis, machine translation – cascaded SMT models). He is the initiator and one of the coordinators of the "COROLA" project on building the representative corpus of contemporary Romanian.

References

1. Atkins, S., Clear J., Ostler, N.: Corpus design criteria. Research report, January 1991
2. Biber, D.: Methodological issues regarding corpus based analyses of linguistic variation. Literary Linguist. Comput. **5**, 257–269 (1990)
3. Biber, D.: Representativeness in corpus design. Literary Linguist. Comput. **8**(4) (1993)
4. Chiarcos, C., McCrae, J., Cimiano, P., Fellbaum, C.: Towards open data for linguistics: linguistic linked data. In: Oltramari, A., et al. (eds.) New Trends of Research in Ontologies and Lexical Resources. Springer-Verlag (2013). doi:10.1007/978-3-642-31782-8_2
5. McEnery, T., Xiao, R., Tono, Y.: Corpus-Based Language Studies: An Advanced Resource Book. Routledge Applied Linguistics, Routledge Taylor & Francis Group (2006)
6. Passonneau, R.J., Ide, N., Su, S., Stuart, J.: Biber Redux: Reconsidering Dimensions of Variation in American English, COLING 2014, pp. 565–576 (2014)
7. Rizzo, C.R.: Getting on with Corpus Compilation: From Theory to Practice. ESP World, vol. 9, issue 1(27) (2010)
8. Sinclair, J.: Corpus creation. In: Candlin, C., McNamara, T. (eds.) Language, Learning and Community. NCELTR Macquairie University, Sydney (1989)
9. Tufiş, D., et al.: CoRoLa starts blooming – an update on the reference corpus of contemporary Romanian language. In: Proceedings of the 3rd Workshop on the Challenges in the Management of Large Corpora, Lancaster, July 2015
10. Tufiş, D., et al.: Reference computational corpus of Contemporary Romanian Language. Research report (in Romanian) June 2015, Romanian Academy (2015)

Practical Session on Text-to-Speech Alignment (Invited Tutorial)

Tiberiu Boroş

Research Institute for Artificial Intelligence "Mihai Drăgănescu",
Romanian Academy
tufis@racai.ro

Abstract. Supervised data-driven speech and natural language processing methods require training data in order to compute their runtime models. The training data has to take the form of input features with associated output and, depending on the application, both the features and output can either be discrete, continuous or mixed. Spoken language processing applications such as speech synthesis and recognition are known for their shared requirement of time-aligned training data, which is not easy to obtain.

During the tutorial, the participants were familiarized with the challenges involved in building time-aligned speech corpora and were presented with a recipe for automatically aligning between sentence-split text and audio, using the well-known flat-start monophones method found in the Hidden Markov Model Toolkit (HTK).

Keywords: Text-to-speech synthesis · Automatic speech recognition · Time-aligned speech corpora · Flat-start monophones

1 Outline of the Tutorial

While speech synthesis can be summed up to learning a transformation from discrete text-based features (syllables, part-of-speech, phonetic transcription, lexical stress, punctuation etc.) to a continuous space signal (described by spectrum, voice pitch and timing), speech recognition has to produce a discrete set of labels starting from the voice signal. These two applications are known to have the shared requirement of time-aligned data which is mandatory in their training phases.

Obtaining training data for speech recognition and speech synthesis is a challenge in itself. While it is possible to obtain prerecorded speech with associated transcripts (news casts, captioned movies, studio recordings etc.), the smallest symbolic representation of speech (which is also used in speech processing) is the phonetic transcription of words and it is likely that the time-alignments associated with the transcripts will not contain this low-level of information. Furthermore, manually aligning phonemes with speech signal requires a tremendous effort, and, due to the exhausting nature of the process, it is prone to error. As such, there has been a lot of research effort invested in the development of techniques that allow bootstrapping speech models and obtaining automatic alignments between speech and phonemes.

One well-known implementation is contained in the Hidden Markov Model Toolkit (HTK) [8] and it is based on the Baum-Welch re-estimation algorithm [2]. One must take into account that before bootstrapping the model, it is preferable that the speech corpus is segmented at sentence level. This can be achieved either manually or addressed automatically (whenever possible) by employing a biased language model speech recognition system [1, 3, 5–7]. Once sentence-level segmentations are available, HTK can be used to obtain fine-tuned phoneme level alignments [4].

2   Short Biography

Tiberiu Boroş is part of the research staff of the Institute for Artificial Intelligence "Mihai Drăgănescu" of the Romanian Academy. He obtained his PhD title in 2013, after the public dissertation of his thesis "Contributions to the modeling and implementation of text-to-speech systems. Case study on Romanian". He participated in several national and EU funded projects, during which the RACAI Text-to-Speech (TTS) synthesis system was developed. Most of his research work is focused on spoken language processing, yielding a number of high-quality contributions which were presented in conferences such as ACL, EANN and LREC. He received a Microsoft Research Award during the Speller Challenge (2011). Currently he is focused in developing natural voice interaction systems for Romanian and his work is centered around building dialogue-based human-computer interfaces and on improving the naturalness of synthetic voices.

References

1. Anumanchipalli, G.K., Prahallad, K., Black, A.W.: Festvox: tools for creation and analyses of large speech corpora. In: WVLSPR, UPenn, Philadelphia, January 2011
2. Baum, L.E., Petrie, T., Soules, G., Weiss, N.: A maximization technique occurring in the statistical analysis of probabilistic functions of Markov chains. Ann. Math. Stat. **41**, 164–171 (1970)
3. Boeffard, O., Charonnat, L., Le Maguer, S., Lolive, D., Vidal, G.: Towards fully automatic annotation of audio books for TTS. In: LREC, pp. 975–980, May 2012
4. Boroş, T., Ştefănescu, D., Ion, R.: Bermuda, a data-driven tool for phonetic transcription of words. In: NLP4ITA Workshop Programme, p. 35 (2012)
5. Charfuelan, M., Steiner, I.: Expressive speech synthesis in MARY TTS using audiobook data and emotionML. In: INTERSPEECH, pp. 1564–1568 (2013)
6. Demuynck, K., Laureys, T.: A comparison of different approaches to automatic speech segmentation. In: Sojka, P., Kopeček, I., Pala, K. (eds.) Text, Speech and Dialogue. LNCS, vol. 2448, pp. 277–284. Springer, Berlin (2002)
7. Stan, A., Bell, P., King, S.: A grapheme-based method for automatic alignment of speech and text data. In: SLT, pp. 286–290, December 2012
8. Young, S., Evermann, G., Gales, M., Hain, T., Kershaw, D., Liu, X.A., Woodland, P.: The HTK book (for HTK version 3.4) (2006)

Contents

Sentiment Analysis in Social Media and Linked Data

Social Data Mining to Create Structured Social Media Resources

Ontological Modeling of Social Media Data

Ontological Modelling of Rumors

Thierry Declerck[1(✉)], Petya Osenova[2], Georgi Georgiev[2],
and Piroska Lendvai[1]

[1] Department of Computational Linguistics and Phonetics,
Saarland University, Saarbrücken, Germany
declerck@dfki.de, piroska.r@gmail.com
[2] Ontotext, Sofia, Bulgaria
{petya.osenova,georgiev}@ontotext.com

Abstract. In this paper, we present on-going work pursued in the context of the Pheme project. There, the detection of rumors in social media is playing a central role in two use cases. In order to be able to store and to query for information on specific types of rumors that can be circulated in such media (but also in "classical" media), we started to build ontological models of rumors, disputed claims, misinformation and veracity. As rumors can be considered as unverified statements, which after a certain time can be classified as either erroneous information or as facts, there is a need to model also the temporal information associated with any statement. As we are dealing in first line with social media, our modelling work should also cover information diffusion networks and user online behavior, which can also help in classifying a statement as a rumor or a fact. We focus in this paper on the core of our rumor ontology.

Keywords: Ontologies · Rumors · Social media

1 Introduction

It is a challenge to decide if statements are in the realm of rumors (including unverified statements, speculation, disputed information, misinformation, and possibly also disinformation) or on the contrary containing facts that can be trusted. The Pheme project[1] focuses on the detection and the classification of rumors across social networks and online media. A task of the project resides in building new and extending existing ontologies to model veracity, misinformation, social and information diffusion networks, rumors, disputed claims and temporal validity.

Ontologies are nowadays widely used as conceptualization models of domains of applications, both in the areas of Knowledge Representation (KR) and Natural Language Processing (NLP). Ontologies act as controlling mechanisms over the relevant data and as means for ensuring adequate inference mechanisms over the stored data. For that reason, in order to represent rumors we rely on the usage of focused ontologies that are modelling the domains of the use cases of the project, but also the types of language data and linguistic phenomena the project is dealing with. For this, we had to design new classes and relations, that could be integrated in existing general purpose

---

[1] http://www.pheme.eu/. See, also [4].

© Springer International Publishing Switzerland 2016
D. Trandabăţ and D. Gîfu (Eds.): EUROLAN 2015, CCIS 588, pp. 3–17, 2016.
DOI: 10.1007/978-3-319-32942-0_1

ontologies. The resulting model we present here is able to represent veracity (including the temporal validity of statements), misinformation and disputed claims on the basis of an annotation scheme that has been developed by sociologists working in close relation with (data) journalists[2]. The ontologies need also to model social and information diffusion networks, users (content authors, receivers and diffusers), lexicalizations and sentiment entities, events and relations. Another goal here is to be able to map and compare extracted statements to data sets published in the Linked Open Data (LOD) framework[3], and more specifically to authoritative sources in the LOD. The ontology is equally named as the project "Pheme", referring here to the figure of the Greek mythology, "an embodiment of fame and notoriety, her favour being notability, her wrath being scandalous rumours"[4].

2   The Pheme Ontology

We decided to develop in a first step an ontology that reflects closely the annotation scheme for social media rumors presented in [9]. It is this version of the ontology that we present in this article. Thus, the presented model focuses exclusively on tweets. It must be noted, however, that it can be easily adjusted towards the incorporation of any further developments of the annotation scheme or towards any other needs we might encounter in its practical use.

Our main strategy was as follows: first, we had to select a top ontology that is suitable to our task. In our case it is the PROTON ontology; then, it was extended by the social media annotation scheme model; finally, it was augmented with other task related ontologies. More detailed descriptions of these steps are given below.

2.1   The Basis: PROTON

PROTON (PROTo ONtology)[5] was selected as the top ontology, also because it supports the linking to DBpedia[6] and other LOD datasets (FreeBase[7], Geonames[8], etc.). The PROTON ontology has been developed in the past SEKT project[9] as light-weight upper-level ontology, serving as a modelling basis across different tasks and domains.

---

[2] See [9]. Some details have been reported in 2014, Deliverable D8.1: Requirements and Use Case Design document. (http://www.pheme.eu/wp-content/uploads/2014/04/PHEME-D8-1-Use-Case-Requirements.pdf).

[3] See http://linkeddata.org/ for more details. Declerck and Lendvai describe in details how to represent social media elements, i.e. hashtags, in the LOD format [3].

[4] Quoted from http://en.wikipedia.org/wiki/Pheme.

[5] http://www.ontotext.com/proton-ontology/. See also [2, 8].

[6] http://dbpedia.org/About.

[7] https://www.freebase.com/.

[8] http://www.geonames.org/.

[9] http://www.sekt-project.com//.

Fig. 1. A partial view on the integration of the Top and Extension modules in the PROTON class hierarchy

PROTON is applied to integrate and access multiple datasets in the FactForge.net[10] – a public service, which allows the user or the application to explore and query efficiently a dataset of 3 billion statements, combining 10 of the most central LOD datasets. The user can either use the original vocabularies of the data sets or the PROTON primitives. In the latter case, the user does not have to deal with the peculiarities of the different datasets. FactForge users benefit also from reasoning on the basis of the PROTON semantics and the owl:sameAs statements made between the datasets.

The PROTON ontology contains more than 500 classes and 200 properties, providing coverage of the general concepts necessary for a wide range of tasks, including semantic annotation, indexing, and retrieval. The ontology is split into two modules: *Top* and *Extension*.

A snapshot of the integration of the Top and Extension modules within the PROTON class hierarchy is displayed in Fig. 1. The top part starts with the prefix *ptop*, which stands for: http://www.ontotext.com/proton/protontop#. The top class InformationResource looks like this:

```
ptop:InformationResource
    rdf:type owl:Class ;
    rdfs:comment "Information Resource denotes an infor-
mation resource with identity, as defined in Dublin Core
(DC2003ISO). Information Resource is considered any com-
munication or message that is delivered or produced, tak-
ing into account the specific intention of its origina-
tor, and also the supposition (and anticipation) for a
particular audience or counter-agent in the process of
communication (i.e. passive or active feed-back)."@en ;
    rdfs:label "Information Resource"@en ;
    rdfs:subClassOf ptop:Statement .
```

The extensions to PROTON have been introduced to cover the conceptual knowledge encoded in Linked Open Data datasets. They start with the prefix *pext*, which means 'the extension of PROTON'.

For example, the extended concept of Artery looks like this:

```
pext:Artery
    rdf:type owl:Class ;
    rdfs:comment "Any artery as a part of the body."@en ;
    rdfs:label "Artery"@en ;
    rdfs:subClassOf pext:BodyPart .
```

---

[10] http://www.ontotext.com/factforge-links/.

2.2   The Pheme Approach

The knowledge in the Pheme ontology has been divided in two levels – *common world knowledge* (including PROTON and different datasets from the LOD) and *Pheme knowledge*, extracted from a set of annotated tweets, describing knowledge about a rumor and its development in time. This dynamic aspect is described in [5]. As mentioned above, the development of the Pheme knowledge module is following the design of the Pheme annotation scheme, which is described in [9] and an application of which is discussed in [6]. Therefore, we followed the division of tweets in *Source* and *Response*, as those have been introduced in the annotation scheme. *Source* and *Response* tweets have different conceptual models with some overlapping features. They share for example the class *Support*.

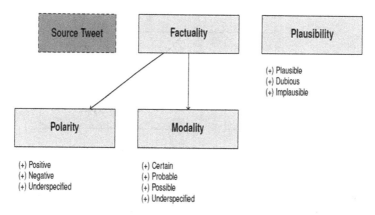

Fig. 2. Annotation scheme model for "Source Tweet" in the Pheme ontology

This annotation is, among others, specifying if a tweet is including elements that are supporting the claim made in the body of the tweet. Below, in Figs. 2 and 3, we display the annotation scheme models for *Source* and *Response* tweets as well as their shared *Support* element (included only in Fig. 3 for avoiding repetition).

Figure 4, below, shows the Pheme classes as modeled in the ontology. We consider a Pheme as a Statement that is expressed in the texts. As a statement, a Pheme has its lifecycle, topic, and truth evaluation. The lifecycle defines its author (Agent), means of creation (Statement), time span (datetime). Topic is defined as a set of RDF statements. Truth evaluation is defined by truthfulness and deception assigned to the topic.

The classes and properties, marked with the prefix ptop, come from the PROTON top ontology (classes: Topic, Statement, Agent; properties: statedBy, validFrom, validUntil). The classes and properties, marked as Pheme, are defined in the Pheme ontology (classes: Pheme, Ptype (= Pheme type), TruthVal; properties: hasTopic, hasType, evaluatedAs, inPost, etc.). The prefix dlpo means that the LivePost ontology[11] is imported into the Pheme ontology. The prefix rdfs

---

[11] See [7] and http://www.semanticdesktop.org/ontologies/2011/10/05/dlpo/.

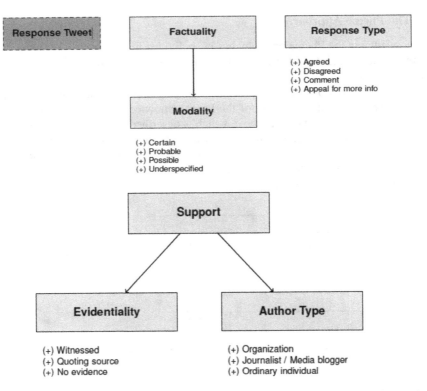

Fig. 3. Annotation scheme model for "Response Tweet" in the Pheme ontology, and the "Support" element, which is shared across "Source" and "Response"

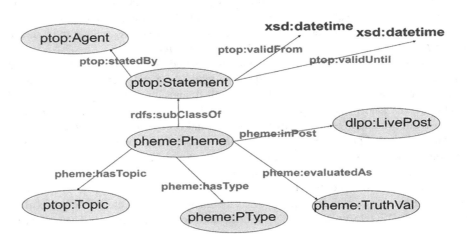

Fig. 4. The classes of the Pheme ontology

means that we are re-using elements from the rdf schema, here to mark that some class is a subclass of another one. The prefix xsd means that some XML schema has been used for typing literal values of datatype properties.

Figure 5 displays subclasses of the type Pheme. For the moment, the unique direct subclass of Pheme is Rumour. Rumour has 4 subclasses: Speculation, Controversy, Misinfor(mation) and Disinfor(mation). The choice of those four subclasses is motivated by the annotation scheme. Here all the presented classes are specific to the Pheme ontology. There is only one relation defined: subClassOf (Fig. 6).

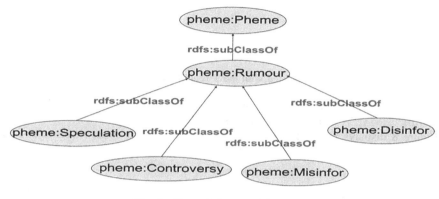

Fig. 5. The subclasses of "Pheme".

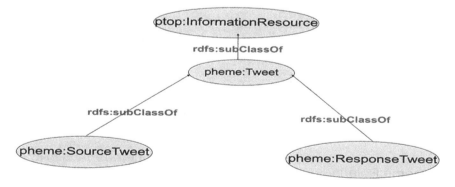

Fig. 6. The relation between ptop: InformationResource and the Source and Response Tweet models of Pheme

Figure 7 shows the 5 properties of the Tweet class. Two of them come from the upper ontology PROTON: hasContributor (Agent) and derivedFromSource (InformationResource); two are specific for Pheme ontology: hasModality (Tweet)

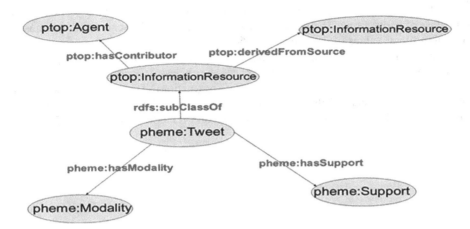

Fig. 7. Properties associated with the Tweet class of the Pheme ontology

and hasSupport (Tweet). The fifth one defines the relation subClassOf with the prefix rdfs: Tweet is a subClassOf InformationResource.

Figure 8 displays the four properties of the SourceTweet class: hasModality, hasSupport, hasPolarity, hasPlausability.

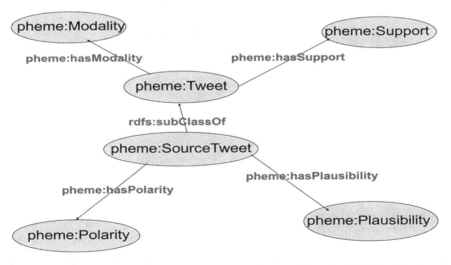

Fig. 8. The properties associated with the SourceTweet class: hasModality, hasSupport, hasPolarity, hasPlausability

And finally Fig. 9 displays the properties of the ResponseTweet class: hasModality, hasSupport (Support), hasRespondType (ResponseType). In this figure only classes and properties that are specific to the Pheme ontology are displayed.

2.3   Example of an Instance of the Pheme Ontology

In the following RDF code, in Turtle syntax[12], we show an example of a source tweet that has been stored as an instance of the Pheme ontology, where the reader can see how some of the classes and relations of our models have been instantiated. The use of the prefix *sioc* refers to the integration of the SIOC (Semantically-Interlinked Online Communities) Core Ontology[13] in our model.

```
pheme:SourceTweet_524947674164760577
         a                           pheme:Rumour ,
pheme:SourceTweet ;
         sioc:has_container
pheme:Thread_524947674164760577 ;
         sioc:has_creator
pheme:UserAccount_DaveBeninger ;
         sioc:topic
pheme:Category_The_soldier_shot_at_War_Memorial_has_died;
         pheme:hasCertainty      pheme:certain ;
         pheme:hasEvidentiality  pheme:source-quoted ;
         pheme:hasSupport        pheme:supporting ;
         pheme:isMisinformation  false ;
<http://www.semanticdesktop.org/ontologies/2011/10/05/dlp
o#textualContent>
                   "BREAKING NEWS: New York Times is re-
porting the Canadian soldier who was shot has died from
their injuries. Heartbreaking. #cdnpoli
#ableg"^^xsd:string.
```

3   Re-use of Existing Ontological Models

We described in some details how we integrated and grounded the Pheme elements in the PROTON top-level ontology. Since we apply our analysis of rumors in the specific field of social media, we also need to propose a model for describing this kind of diffusion of statements. Fortunately we can make use of existing models for this purpose: among those the SIOC and the DLPO ontologies[14]. We present briefly in this section those two models.

---

[12] See http://www.w3.org/TeamSubmission/turtle/ for more details.

[13] See [1] and http://rdfs.org/sioc/spec/.

[14] SIOC stand for "Semantically-Interlinked Online Communities" (see http://rdfs.org/sioc/spec/) and DLPO stand for "Digital.Me LivePost Ontology" (see http://www.semanticdesktop.org/ontologies/2011/10/05/dlpo/).

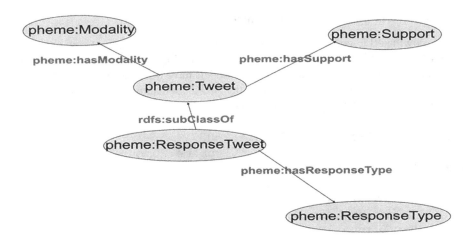

Fig. 9. The properties of the ResponseTweet class

3.1   The SIOC (Semantically-Interlinked Online Communities) Ontology

The SIOC ontology was developed with the purpose of modeling the information that can be published and exchanged in the context of online community sites (weblogs, message boards, wikis, etc.). The motivation being in building bridges between information published in isolated blogs and other social media posts with other sources of information.

And since Pheme is aiming at co-relating distinct sources of information in order for example to support data journalists in their verification work, this model seemed to be very relevant, and we adopted it for our ontological framework. Its re-use was straight-forward, since it is using the same Sematic Web W3C standards that are deployed in PROTON and our Pheme extensions of the top-level ontology, as this is stated in this quotation from the web page of SIOC: "Semantically-Interlinked Online Communities, or SIOC, is an attempt to link online community sites, to use Semantic Web technologies to describe the information that communities have about their structure and contents, and to find related information and new connections between content items and other community objects. SIOC is based around the use of machine-readable information provided by these sites." (http://rdfs.org/sioc/spec/).

Figure 10 below shows the overall organization of the ontological model of SIOC. The reader can see the reasons why we adopted this model: the fact that user groups, user accounts are introduced as classes and that they are explicitly related to the types of post that are being generated in different communities is central to the purposes of verifying if a statement published in social media (or even in classical but online news media). With the adoption of SIOC we are in the position of associating the different types of rumors classified in the Pheme ontology to certain types of users, user groups and also to distinct types of communication platforms. This aspect can help the data journalist in taking different views on statements, depending from whom and from where they originate.

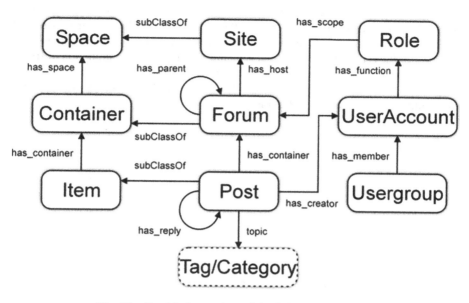

Fig. 10. Graphical overview of the SIOC core ontology

3.2 The DLPO (Digital.Me LivePost Ontology) Ontology

DLPO re-used elements of SIOC but also extends it towards linking between user posts and user presence. A focus of DLPO is to encode the linking of microposts to physical and online user activities and experiences, putting thus more emphasis on personal information of the user and user groups involved in the online communication.

Quoting from the webpage: "DLPO ontology represents the sharing of dynamic, "instant" personal information such as opinions, comments, presence-related information and other multimedia items, as commonly performed across social networks. It represents anything from status messages, to their replies, endorsement (liking, starring), etc. Live Posts refer to online microposts, which can be decomposed in various sub-posts (e.g. posting a video, with a status message, while checking-in at a place and tagging nearby people in the status message). A live post is a subclass of nie:InformationElement and also sioc:Post from the W3C SIOC Member Submission." (http://www.semanticdesktop.org/ ontologies/2011/10/05/dlpo/v1.0/.)

Those elements are very important for Pheme, since the precise encoding of the behavior (frequency and types of activities) of users of social media are important indicators of the degree of factuality expressed in a post. Does a user contribute many types of posts, at what time of the day and in which frequency? All this information could in the end help to distinguish a real user from a robot that reacts automatically to certain key words.

Another very relevant aspect of DLPO is the fact that it integrates a substantial number of other smaller specialized models that have been developed in the context of the Nepomuk project[15] dealing with the development of an integrated suite of ontologies for the establishment of social semantic desktops. Due to limitation of space, we cannot propose a description of those specialized ontologies.

Figure 11 below shows the general architecture of the DLP ontology and lists also some of the integrated specialized ontologies.

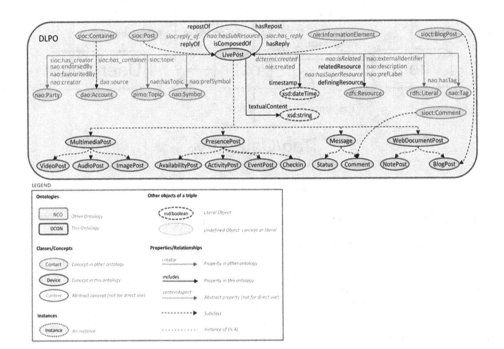

Fig. 11. Visualization of the DLPO ontology

4   Ontology Storage and Maintenance

The integrated PHEME ontology (including the SCIO and the DLPO ontologies), together with the PROTON top ontology has been stored at the Ontotext in-house repository DBGraph: http://pheme-repo.ontotext.com/.

SPARQL queries can be defined over them. For example, the following query can be sent to the Pheme repository:

---

[15] http://www.semanticdesktop.org/ontologies/.

Table 1. Result of a SPARQL query accesing a Replying Tweet

pheme:hasCertainty	pheme:hasCertainty
pheme:hasEvidentiality	pheme:hasCertainty
sioc:has_container	pheme:hasCertainty
sioc:has_creator	pheme:hasCertainty
http://www.semanticdesktop.org/ ontologies/2011/10/05/ dlpo#textualContent	"It's a little scary to know they shoot and kill people suspected o @AntonioFrench" (xsd: string)

```
PREFIX onto: <http://www.ontotext.com/>
prefix pheme: <http://www.pheme.eu/ontology/pheme#>
select * from onto:explicit { ?a a pheme:ReplyingTweet }
```

Table 1 below shows the result of a SPARQL query accessing a Replying Tweet. Here, apart from the Pheme properties (`hasCertainty` and `hasEvidential-ity`), also the thread and creator are pointed out through the integrated SIOC ontology.

If we elaborate further and would like to see also the threads, then the following SPARQL query can be written:

```
PREFIX sioc: <http://rdfs.org/sioc/ns#>
select * { ?a a sioc:Thread }
```

Table 2 below a screenshot is shown, which displays one of the results in all the thread consequence.

Table 2. Result of a SPARQL query concerning a twitter thread

rdf:type	sioc:Thread
rdfs:label	Threads as labelled manually by Pheme partners@en
sioc:container_of	pheme:SourceTweet_524932935137628160
sioc:container_of	pheme:ReplyingTweet_524933104742723584
sioc:container_of	pheme:ReplyingTweet_524933148380237825
sioc:container_of	pheme:ReplyingTweet_524933274444238848
sioc:container_of	pheme:ReplyingTweet_524933385010290688
sioc:container_of	pheme:ReplyingTweet_524933430793553408

5   Conclusion and Current Work

In the figures and examples of codes displayed all along this article we have given a sketch of the current organization of the Pheme ontology, which in this first round of development is mainly reflecting annotation scheme presented by Zubiaga and his colleagues in this year.

We have shown how the Pheme model has been integrated into the PROTON upper model, and how other ontologies (for example the DLPO and the SIOC ontologies) can be imported and integrated in our model. Current work focuses on populating at a large scale the Pheme ontology with data from the automatically annotated tweets.

The creation of the Pheme ontology and its population with correspondingly annotated tweets should finally support LOD-based reasoning about rumors, and this aspect is the focus of the current work. Reasoning with rumors is particularly challenging, due to the need to represent multiple possible truths (e.g. superfoods may cause vs. prevent cancer). The reasoning will be parameterized further in accordance to the domains of the two use cases (healthcare and digital journalism). Reasoning will also take into account the temporal validity of information to accommodate that two otherwise contradictory statements can be both valid at different points in time.

The DBGraph repository, which was mentioned in Sect. 3, will be adapted as a semantic repository with scalable light-weight reasoning. We aim so at supporting efficient inference on tractable dialects of RDF(S) and OWL against billions of facts (RDF triples) that can be accessed in this repository or in the relevant datasets available in the LOD. These datasets provide invaluable large-scale world-knowledge of meronymy, synonymy, antonymy, hypernymy, and functional relations, which are all essential features for classifying entailment and contradictions that can be detected in social media text or in related news documents.

Acknowledgements. This work presented in this paper has been supported by the PHEME FP7 project (grant No. 611233).

References

1. Breslin, J.G., Bojārs, U., Passant, A., Fernández, S., Decker, S.: SIOC: content exchange and semantic interoperability between social networks. In: W3C Workshop on the Future of Social Networking (2009)
2. Damova, M., Kiryakov, A., Simov, K., Petrov, S.: Mapping the central LOD ontologies to PROTON upper-level ontology. In: Ontology Mapping Workshop at ISWC 2010, Shanghai, China (2010)
3. Declerck, T., Lendvai, P.: Towards the representation of hashtags in linguistic linked open data format. In: Vossen, P., Rigau, G., Osenova, P., Simov, K. (eds.) Proceedings of the Second Workshop on Natural Language Processing and Linked Open Data, Hissar, Bulgaria. INCOMA Ltd., Shoumen, September 2015
4. Derczynski, L., Bontcheva, K.: Pheme: veracity in digital social networks. In: Proceedings of the 10th Joint ACL – ISO Workshop on Interoperable Semantic Annotation (ISA) (2014)
5. Lukasik, M., Cohn, T., Bontcheva, K.: Point process modelling of rumour dynamics in social media. In: Proceedings of the 53rd Annual Meeting of the Association for Computational Linguistics and the 7th International Joint Conference on Natural Language Processing (ACL) (2015)
6. Lukasik, M., Cohn, T., Bontcheva, K.: Classifying tweet level judgments of rumours in social media. In: Proceedings of the Conference on Empirical Methods in Natural Language Processing (EMNLP) (2015)

7. Scerri, S., Cortis, K., Rivera, I., Handschuh, S.: Knowledge discovery in distributed social web sharing activities. In: Proceedings of the 2nd Workshop on Making Sense of Microposts (MSM) (2012)

8. Terziev, I., Kiryakov, A., Manov, D.: Base Upper-level Ontology (BULO) Guidance. Deliverable 1.8.1, SEKT project (2005)

9. Zubiaga, A., Liakata, M., Procter, R.N., Bontcheva, K., Tolmie, P.: Crowdsourcing the annotation of rumourous conversations in social media. In: World Wide Web Conference, Florence, Italy (2015)

Towards Creating an Ontology
of Social Media Texts

Andreea Macovei, Oana Gagea, and Diana Trandabăţ[(✉)]

Faculty of Computer Science, "Al. I. Cuza" University of Iasi, Iaşi, Romania
{andreea.gagea,oana.gagea,dtrandabat}@info.uaic.ro

Abstract. Texts live around us just as we live around them. At any instant, there are texts that people write, share, use to get informed, etc. (starting with an advertisement heard on the radio every morning and finishing with the contract of sale signed before a notary). Combining this with the concept of economy in language (or the principle of least effort) – a tendency shared by all humans – consisting in minimizing the amount of effort necessary to achieve the maximum result, it is no wonder why the social media, with its short, informal and context dependent texts, achieved such a high popularity.

Even texts are so constantly present in our lives (or precisely because of that), linguistic classification of texts is still debated, and no clear visualization of texts types is yet available. Going beyond the classification of texts in species and genres, this paper proposes an ontology which discusses the various text types, focusing on social media texts, and offering a set of properties to describe them.

Keywords: Social media analysis · Ontology · Text mining

1 Introduction

Slowly but surely, social media replaced the traditional sources of information: people's need to be constantly updated changed our behavior from buying a newspaper or watching TV, to using a Facebook or Twitter account to visualize, in a customizable manner, the day's hottest news, with the bonus of being able to also comment on them.

Also, texts shared through social media applications offers us the information that we need: for example, the reviews of a product are texts that provide us useful information regarding that product, the text of recipe exposes us the steps that we must follow in order to prepare a cake, while the text of an advertisement invites us to eat at the new Chinese restaurant in town.

As huge amounts of texts become available through social media, a challenging task concerns the organization and processing of this information to extract knowledge. Natural language processing tools trained on large news corpora have usually problems when applied to unstandardized social media inputs, firstly due to the fact that social media content can appear in various forms [3], from photos and video updates to news, offers and literary works, and secondly since, on regular basis, social networking texts are written in informal, shortened and context dependent style.

© Springer International Publishing Switzerland 2016
D. Trandabăţ and D. Gîfu (Eds.): EUROLAN 2015, CCIS 588, pp. 18–31, 2016.
DOI: 10.1007/978-3-319-32942-0_2

This paper introduces an ontology of social media texts types, along with their properties and relations, intended to be further used to automatically tag social media texts with their type, to facilitate the selection of proper information extraction tools.

A common definition of ontology considers it to be a "specification of a conceptualization" [10], where conceptualization is an abstract view of the world represented as a set of objects. The term has been used in different research areas, including philosophy, artificial intelligence, information sciences, knowledge representative, object modeling, and most recently, eCommerce applications. For our purposes, we follow the definition in [16] and consider ontology to be a directed graph, with nodes representing concepts associated with certain semantics (properties), allowing to specific relations between them.

The remaining of this paper is structured as follow: Sect. 2 shortly presents the current research in joining ontologies and semantic media analysis; Sect. 3 shortly introduces an upper ontology of literary and non-literary genres and species; Sect. 4 discusses some examples from social media texts, and Sect. 5 draws the conclusions and directions for further work.

2 Social Networks and Ontologies

Inspired by the most known ontologies of wine and food [6, 13], the task of developing an ontology for text types was necessary in order to adjust the processing tool for a specific task (i.e. POS tagging) to the processed text type[1]. Offering a set of properties and relationships between text types, our ontology can also be used to create suggestions for readers, to identify a certain type of text or to analyze the format differences between two similar types of texts (as offers and advertisements, for instance), all as a first step in automatic text writing.

In the last decade, researchers have started to consider different approaches and applications to merge the social media and semantic web domains. Rainer and Wittwer [15] discuss existing approaches which combine text mining techniques with ontology creation for social media texts. Most ontology extraction models which use social networks, such as the one described in [11], rely on three dimensions: actors, concepts and instances, and illustrates ontology emergence using actor-concept and concept-instance relations.

Stan et al. in [17] use social networking to extract a user profile ontology that allows to efficiently characterize the current situation of the user and to express social-network related reachability preferences in a set of situational sub-profiles. Different ontological analysis of social media are applied to building emerging social networks by analyzing the individual motivations and preferences of users, broken into potentially different areas of personal interest, such as the one in [5].

---

[1] For best results, the processing tool needs to have been trained on the same type of texts as the text to be processed.

Crisis management and early detection of possible crisis situations (such as disasters, conflicts and health threats) from a continuous flow of on-line news articles are favorite applications of social media analysis using ontological information [14, 18].

Somehow more related to our approach is Bruce Schneier's [4] taxonomy of social networking data, identifying a set of different data types which characterize social networking: *service data* (data gave by social network users), *disclosed data* (posts on a user's page), *entrusted data* (posts on other people's pages by a specific user), *incidental data* (posts by other people about a specific user), *behavioral data* (data collected by the site) and *derived data* (data derived by analyzing the social network of a user). In the same line of research, [8] proposes a hierarchically-structured taxonomy for online social networks data types by studying fundamental user activities on social networks and step-wise classifying identified data types into non-redundant partitions.

While we agree with [12] that it is extraordinary to see semantics emerging from individual actions of a community, the main purpose of this paper is yet different. Thus, the major difference between our approach and existing methods is that, while the above presented solutions rely on the content of social media texts, our proposed classification primary considers the form rather than the fond of texts in social networks.

3   Upper Ontology of Text Types

The semantic media ontology is part of a larger project, an ontology of text genres, both literary and non-literary, establishing hierarchies and connections between texts types. The aim of this paper is not to discuss the classification of literary genres and sub-genres, since the authors acknowledge that the question of literary genres has been disputed for centuries, and are well aware of the limitations of different taxonomies. In this context, the upper ontology of text genres we proposed is rather intended to provide a set of properties and examples for different species in each genre. Two main text types are considered: literary genre and non-literary genre, to which a special category was added: social media texts (to be detailed in the next section).

3.1   Literary Texts

Theoretical studies of Romanian literature refer especially to one text genre: *literary genre*, known also as *fiction literary genre*. Aristotle himself divided literary texts into superior drama or tragedy, inferior drama or comedy, superior narrative or epic and inferior narrative or parody. Starting with Aristotle and finishing with the literary theorists of these days, the classification of text types is still debated.

On the other hand, no specifications about other genres in Romanian literature were found: literary critics [1] present the idea of *non-literature* but they do not provide a precise classification of species that could belong to this category.

So, unfortunately, a predefined list of text genres does not exist as there are numerous disagreements regarding the definition of specific genres: a genre for a theorist could be a sub-genre or even a super-genre for other theorists [7]. But, there is an actual and clear distinction between *literary texts* and *non-literary texts* in Romanian [9]:

- A *literary text* is a text by which the author wants to impress and to enthuse readers, expressing his or her own thoughts, ideas and feelings, using an artistic language, strongly marked by subjectivity.
- The *non-literary text* is a text which aims to objectively inform the reader about certain aspects of reality in a clear and precise language, often containing scientific and technical terms.

The class of literary texts includes the literary works, divided into four main categories: (1) epic genre, (2) lyrical genre, (3) drama genre and (4) argumentative genre, each genre having its own species. Table 1 below presents the main sub-classes of genres included in the literary class, as well as examples of species for each genre.

Table 1. Species of literary texts

Ontology of text genres	
Literary class	Examples of text types
Epic genre	Diary, Biography, Autobiography, Memories, Epic, Novel, Sketch story, Short story, Novella, Fairytale, Myth, Parable, Ballad, Fable, Poem, Anecdote
Lyric genre	Elegy, Ode, Pastel, Meditative poetry, Satire, Pamphlet, Sonnet, Rondo, Ghazel, Gloss, Romance, Hymn, Doina, Idyll, Haiku
Drama genre	Drama, Comedy, Tragicomedy, Tragedy, Theatre of the absurd
Argumentative genre	Argumentative, persuasive texts

3.2 Non-literary Texts

Regarding the non-literary class of texts, we consider five categories: (1) informative genre, (2) instructive genre, (3) persuasive genre, (4) juridical genre and (5) descriptive genre.

An *informative text* is the text informing the reader about the natural or social world, while *an instructive text* is a text that instructs or tells the reader how to do something. *A persuasive text* is a text used to convince a reader about a particular idea or focus. The *juridicial genre* is represented by laws, contracts, regulations and the *descriptive genre* by descriptive texts as those found in travel guides.

For these types of text, we identified a set of lexical, syntactic and stylistic features which can be used to automatically classify a text into one of these categories, if any.

The main difference between literary and non-literary genre consists in the subjective/objective character of texts: the involvement of the author characterizes literary works, while usually non-literary texts lack direct author involvement. There are, of course, exceptions, such as offers, advertisements, reviews, social media content, scientific and descriptive texts (i.e. brochures), which can bear (direct or hidden) involvement of the author, even if they fall in the non-literary class. Other differences may be found between various species from the two main classes (literary and non-literary), such as the ones related to the length or to specialized language of the

Table 2. Species of non-literary texts

Ontology of text genres	
Non-literary class	Text types
Informative genre	Weather reports, Reportage, Press article, News, Reviews, (Instruction), Scientific texts
Instructive genre	Manuals, Recipes, How-to guides texts
Persuasive genre	Offers, Advertisements texts
Descriptive genre	Travel guide, Descriptive texts
Juridical genre	Contracts and dispositions, Law texts, Regulations

texts, but they may not be generalized. Table 2 presents the species included in each sub-class of genres from the non-literary texts.

3.3   Properties and Relations

A set of properties was developed to characterize the text types in our ontology. Each text type is given specific values of the considered set of properties. For instance, we have the commercial feature of several text types, such as *offers* and *advertisements*. Their commercial character is identified through the analysis of the structure of the texts, but also from morphologic and syntactic information (first person of the verbs, exclamations and indications as address or telephone number).

The set of properties refer to the functional style, the presence/absence of spatial and temporal indications, the mood of the transmitted message, presence/absence of figures of style, specialized or general words, presence of a dominant entity (more precisely, if the texts are centered on a single entity or not), the length of the text, prosody, commercial style etc. Table 3 presents the set of features used to describe each species in the ontology (second column) and the class of their associated values (third column).

The true/false values are not treated in a classical, binary, manner, but in a fuzzy way. The values for true range from 1 to 4, according to how frequent the specific structure appears in the considered species (1 - less frequent, 2 - relatively frequent, 3 - more frequent and 4 - very frequent).

A graphical ontology editor (OWLGrEd) is used in order to provide a visual overview of all the categories of texts that can be found and to establish relations between different types of texts. The relations we considered so far are: IS-A, *disjoint of, complement of, instance of*. The next section presents the part of our upper ontology dealing with social media text types.

Table 3. Taxonomy of properties used to describe text types in our upper ontlogy

Types of features	Features	Value set of feature
Structural features	Format (fixed form)	true/false
	Length	short/medium/long text
	Connectors	true/false
Lexical and morphological features	Imperative mood and vocative case	true {values 1|2|3|4}/false
	General (common) or Specialized words	general/specialized language
	Spatial and temporal indications	true {values 1|2|3|4}/false
	Presence of entities (texts centered on one or more entities)	true {values 1|2|3|4}/false
Stylistic features	Prosody (rhyme, verse, strophe)	true/false
	Figures of style	true {values 1|2|3|4}/false
Semantic features	Functional style	colloquial/belletristic/publicistic/scientific/administrative style
	Commercial style	true/false
	Mood	optimistic/pessimistic/neutral mood
	Involvement of the author	direct/indirect involvment

4   Ontology of Social Network Text Types

Texts that can be found on social media applications are as various as possible: ranging from review, blogs and forums to social networking sites as Twitter and Facebook. The latter are preferred by users, as they stay informed and connected to their peers using profile accounts. Given the variety of users and generated contents, is it possible to classify texts of social media data according to the proposed ontology of texts?

Figure 1 shows a classification of social media content according to what users usually post and share through social media applications.

Specific processing tools (such as POS taggers or anaphora resolution systems), score a higher performance if used on the same text type as the ones they were trained

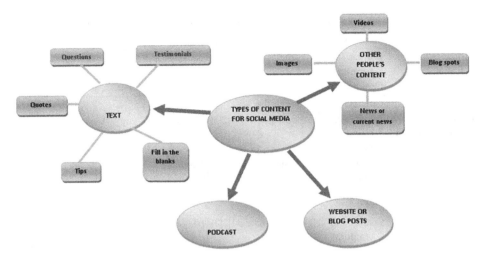

Fig. 1. Types of posts for social media content

on. In other words, we will have better results if using a POS tagger trained on news corpora to analyze news texts, rather than speech transcripts.

Although social media texts are usually short[2], we found out that they may have a variety of formats. In this context, a classification of social media text types becomes important if we want to extract reliable knowledge from social data.

To our best knowledge, there exists nowadays no ontology for social media text types, nor its extension for all texts genres. Therefore, starting from our upper ontology for text genres, we developed an ontology of social networking data, containing properties for each type of texts, examples and relations between texts.

Social media content found on Facebook or Twitter pages can be divided into several categories: personal data, informative content (such as news or weather reports), instructive data (such as manuals, reviews, how-to guides – in all multimedia format types), persuasive data (such as advertisements or offers), descriptive content (mostly in the form of links to travel guides) and literary works (anecdotes, poems, short stories etc.). Figure 2 presents the types of social media text considered in this research.

As one can see, these types belong to different categories from the upper ontology, both from literary and non-literary classes. Each text type bears its own set of features, as the one presented in Sect. 3.3, with specific values.

A new category, not introduced in the upper ontology, is Personal Data. It refers to all the information provided by the users about themselves or regarding other people and events (videos, status updates, spatial or temporal indications, photos, public or private postings, etc.). Besides personal data, social media is a favorable environment

---

[2] Longer social media content can also be found, usually introduced as a link.

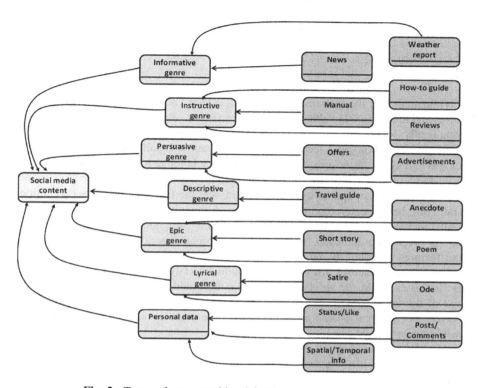

Fig. 2. Types of texts considered for the social media ontology

where different types of content can be collected: these text types can be similar with other traditional forms of publishing [2].

The types of content identified on social media have specific features that can be used for an automatic identification of each type of text. However, there are also similarities between those types: for example, offers, advertisements and reviews are always shorter than weather reports or literary works. The text of a weather report contains specific words as *meteorological conditions, the northern half of a country, wind intensity*, etc. It is not the case for reviews, offers, advertisements where the style of language is colloquial in order to make all the people to understand the message.

The following examples, extracted from Romanian Facebook and Twitter pages, show how different in both content and form can Facebook and Twitter pages posts be.

Example 1. Poem[3]:

RO.: Capriciu/ Aş vrea să fiu/ Un capriciu de-al tău/ Sau un moft/ Şi să mă înveţi din asta/ Pe de rost,/ Să mă citeşti/ Ca pe o carte deschisă/ Şi să mă joci,/ Cap sau pajură,/ Ca pe-o iubire promisă.

---

[3] The poem can be found on: https://www.facebook.com/mariustuca.ro?fref=nf.

En.: Caprice/I want to be/ A caprice of yours/ Or a whim/ And you learn from it / By heart,/ How to read me/ Like an open book/And how to play with me,/ Heads or tails,/ Like a promised love.

While analyzing the features for this example (see Table 3 above for the full list of features and their possible values), we can observe that structural features are easy to be automatically identified: no fixed format, short length, abounding in connectors. Lexical and morphological features can also be automatically identified using morphologic and syntactic information: the vocabulary belongs to a general one, with no specialized terms; the conjunctive verbal time indirectly indicates an imperative mood, there is no mention of spatial or temporal data, and the fragment presents two entities (I and You), being centered on the second one. Although more difficult to identify, the stylistic features can also be occasionally automatically identified: the presence of "like" as preposition indicates the presence of comparisons as figures of style, and the poem features a broken rhyme in some of its verses.

The semantic features are the most difficult to analyze, requiring a trained annotator. In this example, the involvement of the author (and the reader) is indicated by the usage of the first and second person of the verbs and there is no commercial aspect involved, but the real challenge is to identify the overall mood of this short poem, possibly an optimistic one.

As species belonging to non-literary texts, the following examples should be characterized by objectivity and marks of indirect involvement of author. These two indicators appear in texts in various forms, which can be automatically identified: impersonal verbs, the third person of verbs, the third person of pronouns, lack of adjectives, figures of speech, prosody aspects and description markers, declarative phrases and exclamations, connectors, etc. It is important to see if these general features applying for non-literary texts characterize or not the proposed posts and if there are other features that are specific for a certain text type.

Example 2. Advertisement[4]:

Ro.: Lodgy Stepway este partenerul de călătorie ideal pentru weekend-ul prelungit, nu credeți? Descoperiți mai multe despre el!

En.: Lodgy Stepway is the ideal travel partner for the next weekend, don't you think? Discover more about it!

Analyzing the structural features of this example, it shows no fixed format, a short length and not so many connectors as in the first example. Lexical and morphological features indicate a general vocabulary (needed to make the advertisement reach a wide category of potential clients), no specialized terms, and the use of imperative mood (*discover*), rhetorical questions and temporality (*now*) should convince the reader to find out more. Only one entity (the name of the car) is presented, and as stylistic

---

[4] This advertisement can be found on: https://www.facebook.com/DaciaRomania/photos/a.501429966539492.132624.204829606199531/1082250568457426/.

features, this advertisement, as most of them do, personalizes the presented object, to make it look as a perfect partner. The involvement of the author and the reader is obvious by the usage of the structure *don't you think* and the second person of the verb, and the overall mood of this example is clearly optimistic.

The offers and advertisements which appear in online environment are the only types of texts featuring a commercial character, expressed by the structure of texts: the presence of exclamation marks, capital letters, repeated words, rhetorical questions, abundance of adjectives, usually in superlative, all representing important indicators to be used as patterns for automatically identifying this type of social media content.

But, there are other shared texts that seem to be persuasive despite offers and advertisements:

Example 3. Persuasive text[5]:

Ro.: De ce se îmbracă oamenii de succes mereu la fel: care este explicația

En.: Why do successful people always dress the same? What explanation can there be?

For the Example 3, the author uses a question (*Why do successful people always dress the same?*) in order to draw the attention of readers: questions are not a general feature that may indicate a non-literary text, but in case of persuasive texts, it can be a distinctive mark. As an answer of the question, another question arises: the adverb *there* and the verb *to be* strengthen the beliefs of reader and make him or her read the answer. In this case, rhetorical questions must be added as features specific for the class of persuasive texts.

Searching for other similar examples of persuasive texts, it can be observed that interrogations predominate in Twitter messages:

Example 4. Persuasive text:

Ro.: Cât te costă vacanța de Crăciun și de Revelion în Poiana Brașov și pe Valea Prahovei

En.: How much does it cost Christmas and New Year holiday in Poiana Brasov and Prahova Valley?

Example 5. Persuasive text:

Ro.: Cum a reușit clanul Cârpaci să primească un cadou de jumă-tate de milion de euro de la CNADNR?

En.: How did the clan of Cârpaci manage to receive a gift of half a million euros from CNADNR?

The example below is a book review, an one sentence content with no fixed format, a short length and no connectors at all. The review uses a general vocabulary, and tries to be as objective as possible: no imperative mood, no reference to spatial/temporal indicators, no focused entity (the review equally promoting both the author and the book), no prosody, no figures of style, no commercial style, and a neutral overall mood.

---

[5] The post is available on: https://twitter.com/adevarul/status/653595042523807744?lang=bn.

The only feature characterizing this example is the involvement of the author, made obvious by the usage of appreciating adjectives.

Example 6. Review[6]:

Ro.: O autoare apreciata, un roman pretuit de critici si de publicul larg: Ne-am iesit cu totii complet din minti, de Karen Joy Fowler

En.: An appreciated author, a novel cherished by critics and general public: We Are All Completely Beside Ourselves by Karen Joy Fowler

As reviews are considered to belong to instructive genre as manuals and how-to guide, here are some examples of instructive posts which cannot be integrated in these subclasses:

Example 7. Instructive text:

Ro.: Adevărul din spatele fotografiei de nuntă perfecte. Ce sacrificii fac fotografii în timpul şedinţei

En.: The truth behind the perfect wedding pictures. What sacrifices take photographs during the photo session?

The example above is included in the class of instructive texts because first of all, it exposes a situation without using a verb (*The truth behind the perfect wedding pictures*). The text continues with an interrogation (*What sacrifices take photographs during the photo session?*) where the word *sacrifices* can resume a number of tasks that photographs have to deal with. Indications regarding a possible enumeration (or a list – plural number of the noun *sacrifices*) or several reasons (presence of numbers and of modal verbs in Example 8) can be objective indicators emphasizing an instructive text.

Example 8. Instructive text:

Ro.: 5 motive pentru care merită să stăm cu ochii pe Simona #Halep

En.: 5 reasons why we should keep an eye on Simona #Halep

The class of informative genres is represented by news and weather reports as it is showed in the following examples:

Example 9. Weather report:

Ro.: VREMEA în weekend: Prognoza meteo în ţară, la Bucureşti, la mare şi la munte: Sâmbătă, VREMEA va fi în general...

En.: Weather this weekend: Weather forecast for country, Bucharest, seaside and mountains: On Saturday, the weather will be generally...

---

[6] The review is available on: https://www.facebook.com/groups/cartidecitit/.

This type of informative text contains many temporal (*week-end, Saturday*) and spatial indications (*country, Bucharest, seaside, mountains*). A repetition of the term *weather* can be also an indicator of informative texts. Titles of press articles and news (Example 10) are the most frequent Twitter informative texts.

Example 10. News:

Ro.: `Tribunalul Bucureşti a amânat discutarea dosarului privind falimentul Astra Asigurări`

En.: `Bucharest Court postponed the discussion on the case of the Astra Insurance bankruptcy`

Even if these types of texts belong to the same class (literary or non-literary texts), they are different. This means that general features of literary and non-literary texts do not always match for each text (for example, the third person of verbs or pronouns); and it may happen that non-literary texts "borrow" some features from literary texts (the first and second person of pronouns and verbs, in these cases). The features that have been identified in texts for each species provide sufficient information in order to establish patterns to be used for the program that automatically classifies those posts and tweets.

If we consider the non-literary texts, our examples (all the texts except for Example 1) should be characterized by objectivity and marks of indirect involvement of author. These two indicators appear in texts in various forms, which can be automatically identified: impersonal verbs, the third person of verbs, the third person of pronouns, lack of adjectives, figures of speech, prosody aspects and description markers, declarative phrases and exclamations, connectors, etc.

But, this is not always the case, as we have showed: it is important to see if these general features applying for literary and non-literary texts characterize or not the proposed examples and if there are other features that are specific for a certain text type.

The large amount of texts belonging to different genres becomes more useful if it is organized. The presented ontology can become a way to represent the textual variety. If the types of texts are determined and their features and examples are used as indicators for automatically delimitating other similar types of content, the ontology can become a useful tool for developing natural language processing applications.

5   Conclusions

As various types of texts are shared every day, a method of structuring those texts can be very necessary. If it is possible to determine the type of text, many applications can be performed in order to extract information, to analyse the content of the text, to offer suggestions, to summarize a text, to identify a text according to its format, etc. Here are several use cases of such ontologies:

– a ontology can be useful in order to classify texts: domain corpora can be created in an automatic manner;
– extracting information from a literary or non-literary source in order to be further processed will be possible;

- programs that offer reading suggestions for people can be developed;
- if there are specific formats of texts (as in the case of *law texts*), an application that identifies a type of text can be built;
- a clear and exact text typology can help specialists to determine the particularities and similarities of a type of text; these observations can be used in order to automatically identify a certain text;
- in one way, a genre ontology can propose an overview of all texts existing in a language: this can represent the starting point for various ontologies of different languages and comparative studies between two or more languages in terms of literature.

As social networking has become a must in our society, and taking advantage of this volunteer interaction to extract knowledge is an asset most data researchers value. However, social data comes in a wide range of text types, from news, to poems or speech transcripts, most of them needing adequate, specially trained processing tools.

The proposed ontology of social media text types is a useful instrument to classify different types of social networking texts. Starting with the proposed text types, we built an ontology by identifying different properties and relationship between these text classes.

Analyzing different types of texts in social media, we observed a variety of forms and contents, which is explained by the wide diversity of content creators and intended readers. While some features are vary largely, depending on the type of texts, other features, such as limited prosody, high involvement of the authors and expected involvement of the reader, generalized vocabulary, free format, etc. tend to be similar across contents.

An immediate possible application of using the proposed social media ontology is to find and recommend content similar to the text types offered by the user, or to filter specific text types (such as advertisements or jokes).

Also, automatic classification of social media content according to specific types of text could illustrate the diversity of texts shared and posted through social applications. At the same time, it could be a premise in detecting the style of each Twitter or Facebook user, the creativity of his or her posts and the level of involvement in order to draw the attention of virtual public.

References

1. Bahtin, M.M., Iliescu, N., Vasile, M.: Probleme de literatură şi estetică. Univers, Bucureşti (1982)
2. Barwick, K., Joseph, M., Paris, C., Wan, S.: Hunters and collectors: seeking social media content for cultural heritage collections. In: Proceedings of VALA2014 17th Biennial Conference, Melbourne (2014). http://www.sl.nsw.gov.au/about/publications/docs/BarwickJoseph-VALA.pdf
3. Becker, H., Iter, D., Naaman, M., Gravano, L.: Identifying content for planned events across social media sites. In: Proceedings of the Fifth ACM International Conference on Web Search and Data Mining, pp. 533–542. ACM, New York (2012)

4. Schneier, B.: A taxonomy of social networking data. IEEE Secur. Priv. **8**(4), 88 (2010). doi:10.1109/MSP.2010.118

5. Cantador, I., Castells, P.: Building emergent social networks and group profiles by semantic user preference clustering. In: 2nd International Workshop on Semantic Network Analysis (SNA 2006), at the 3rd European Semantic Web Conference (ESWC 2006) (2006)

6. Cantais, J., Dominguez, D., Gigante, V., Laera, L., Tamma, V.: An example of food ontology for diabetes control. In: ISWC 2005 Workshop on Ontology Patterns for the Semantic Web, Ireland. http://citeseerx.ist.psu.edu/viewdoc/download?doi=10.1.1.137. 8373&rep=rep1&type=pdf

7. Chandler, D.: An introduction to genre theory. Media and Communication Studies (1997). http://www.aber.ac.uk/media/Documents/intgenre/intgenre1.html

8. Richthammer, C., Netter, M., Riesner, M., Pernul, G.: Taxonomy for social network data types from the viewpoint of privacy and user control. In: ARES 2013, 2013 Eighth International Conference on Availability, Reliability and Security (ARES), pp. 141–150 (2013). doi:10.1109/ARES.2013.18

9. Enciu, V.: Introducere în teoria literaturii, Curs universitar, Bălţi, Presa universitară bălţeană (2011)

10. Gruber, T.R.: A translation approach to portable ontology specifications. Knowl. Acquisition **5**, 199–220 (1993)

11. Hamasaki, M., et al.: Ontology extraction using social network. In: International Workshop on Semantic Web for Collaborative Knowledge Acquisition (2007)

12. Mika, P.: Ontologies are us: a unified model of social networks and semantics. In: Gil, Y., Motta, E., Benjamins, V., Musen, M.A. (eds.) ISWC 2005. LNCS, vol. 3729, pp. 522–536. Springer, Heidelberg (2005)

13. Noy, N.F., McGuinness, D.L.: Ontology engineering for the semantic web and beyond. Stanford Medical Informatics, Stanford University (2000). http://protege.stanford.edu/publications/ontology_development/ontology101.html. Accessed 11 Jan 2005

14. Pröll, B., Retschitzegger, W., Schwinger, W., Kapsammer, E., Mitsch, S., Baumgartner, N., Rossi, G., Czech, G., Högl, J.: CrowdSA - crowdsourced situation awareness for crisis management. In: Proceedings of the International Conference on Social Media and Semantic Technologies in Emergency Response (SMERST), Coventry, UK, April 2013

15. Rainer, A., Wittwer, M.: Towards an ontology-based approach for social media analysis. In: Proceedings of ECIS (2014)

16. Russell, S., Norvig, P.: A Modern Approach: Artificial Intelligence, 2nd edn. Pearson Education, Inc., Egnlewood Cliffs (2003). ISBN 0-13-790395-2

17. Stan, J., et al.: A user profile ontology for situation-aware social networking. In: 3rd Workshop on Artificial Intelligence Techniques for Ambient Intelligence (AITAmI2008) (2008)

18. Vanni, Z., Tanev, H., Steinberger, R., van der Goot, E.: An ontology-based approach to social media mining for crisis management. In: Proceedings of the Workshop on Social Media and Linked Data for Emergency Response (SMILE'2014), Anissaras, Crete, Greece, 25–29 May 2014 (2014)

Application of Social Media and Linked Data Methodologies in Real-Life Scenarios

Towards Social Data Analytics for Smart Tourism: A Network Science Perspective

Alex Becheru, Costin Bădică$^{(\boxtimes)}$, and Mihăiță Antonie

Computers and Information Technology Department, University of Craiova, Craiova, Romania
becheru@gmail.com, cbadica@software.ucv.ro,
mihai.antonie@gmail.com

Abstract. In this paper we present our preliminary results regarding collecting, processing and visualizing relations between the user comments that were posted on Smart Tourism Web sites. The focus of this paper is on investigating the user interactions generated by expressing questions and answers containing the users' impressions and opinions about the attractions offered by various tourism destinations. We propose a prototype system based on the design of a conceptual data model and of the development of a data processing workflow that allows to capture, to analyze and to query the implicit social network that was determined by the relations between user comments, using specialized software tools for graph databases and complex networks analytics.

Keywords: XML processing · Complex network · Graph database · Smart tourism

1 Introduction

A great interest was shown during the last decade to the application of information and communication technology (ICT) for the development of Smart Tourism business [10], here understood as "the application of information and communication technologies to the tourism sector"[1]. Tourists are interested to benefit from the availability of advanced ICT services for knowledge and information management to assist them in taking informed decisions matching their preferences with less effort and in shorter time. Tourism companies look for improving the quality of their services, as well as their image, based on the feedback gathered from their customers.

There are many information sources assisting tourists for their consumption needs, including personal and company, as well as general and specialized Web sites and social media applications [17, 20]. They can be defined as content sharing networks that allow their customers to share and exploit reviews and opinions about tourist destinations, including: post-visit experiences, tourist advertisements, descriptions of tourist attractions, tourist highlights and advices, recommendations, and geo-tagged photos, addressing various aspects including accommodation, trips, historical places, landscape, sightseeing, food, restaurants, shopping, entertainment, local attractions, weather, a.o.

---

[1] http://www.smarttourism.org/.

© Springer International Publishing Switzerland 2016
D. Trandabăţ and D. Gîfu (Eds.): EUROLAN 2015, CCIS 588, pp. 35–48, 2016.
DOI: 10.1007/978-3-319-32942-0_3

Tourist information is most often presented as reviews or comments expressed using natural language text that describes customer opinions or experiences about various tourist entities. The users have also the possibility to interact by: (i) providing echoes, as well as answers to echoes posted in relation to certain reviews or comments, and (ii) asking questions and giving answers to questions about a certain tourist place. This process follows forum-like interaction patterns and triggers dialogues between users that define implicit social networks of people that share interests on similar or related tourist topics. Collecting, aggregating and presenting this information in a meaningful way poses cognitive challenges to the users, as well as technical challenges to the computational methods employed.

The paper is focused on the automated analysis of data gathered from tourism Web sites by exploiting the implicit network representation of this data. Our results include:

- A method for reconstructing the implicit social network determined by the users' discussion about various tourism entities.
- The application of network analysis tools to the resulted social network.
- The representation and querying of the social network using graph database tools.

The paper is structured as follows. Section 2 presents some background information on network science. The next part of the paper enumerates some related works. In Sect. 4 we present the architecture of our system, as well as the technical details regarding data gathering and preprocessing. We introduce the conceptual model of our data, discussing the technical challenges of data preparation for graph-based data processing and analysis. In Sect. 5 we consider to use cases for our system: data querying using a graph-based database management system and graph-based analysis using a network analytics tool. The last section presents the conclusions and future work.

2   Network Science Background

Network science or Complex Networks Analysis (CNA) is based on graph theory. The aim of this research field is to address networks with non-trivial features, features that do not occur in lattices or random graphs. Networks with similar properties can be found in many networks extracted from real life phenomena. The complexity of a real world network comes from understanding and evaluating overlapping phenomena that are neither purely regular nor purely random. Also complexity may come with the vast size of the underlying network itself.

Two papers stand as the building blocks of Network Science. In 1959 Paul Erdős and Alfréd Rényi published a study on random networks as part of graph theory [13]. Mark Granovetter wrote in 1973 about the strength of the weak ties [16], ties which hold together communities that are otherwise divergent. The necessary conditions for the creation of Network Science were fulfilled in the early 1990. At that point the technological evolution reached a state where both the computational power and storage necessary for analyzing Complex Networks became affordable. Access to real world networks was facilitated by the invention of the World Wide Web together with an explosion of detailed map-making across many fields of research.

In the business environment *Google's search engine* and *Facebook* are based on complex networks. Internet search and social networks are based on network topology and characteristics. In medicine *Network Science* has been used to understand the spread of diseases and to develop effective methods to stop the escalation of such diseases on larger scale [6]. Security forces use CNA to discover knowledge about wanted people that can lead to their finding/arrest, e.g. the arrest of Iraq's former dictator *Saddam Hussein* [24].

3 Related Works

The recent research literature contains many works related to the generous subject of Smart Tourism. In particular, a lot of papers are focused on recommender systems, trying to valorize the rich informational context of this domain. Fewer papers are proposing the use of network analysis tools for exploring tourism information sources.

Paper [23] investigates the role, level, as well as gap and opportunities of Web 2.0 for empowering knowledge management of tourism companies in Greece. The study revealed that although many Greek tourism professionals use content sharing networks, their greatest majority did not exploit this opportunity for knowledge management activities. Our work can be seen as an attempt for narrowing the gap between importance and utilization of shared content for knowledge management. A related paper focused on using social media for travel related search is [11].

A lot of works are aimed to develop new recommender systems for tourism that exploit the recent advances in ICT, including mobile computing [15, 21, 25], semantic technologies [2, 14], and geo-tagged information [18]. A recent survey of tourism recommender systems is presented in [9].

Three recent papers that are focused on the analysis of data extracted from AmFostA-colo are [4, 7, 12]. Paper [4] introduces a sentiment classification method for the categorization of tourist reviews according to the sentiment expressed in three categories: positive, negative or neutral. Paper [12] presents results of sentiment analysis for relating the tourist opinion holder with the review content. Paper [7] explores complex network representation of tourist reviews for extracting lexical and quantitative features of the review text.

Other works employing network analysis computational methods in the tourism domain [3] are and [22]. Paper [3] employs network theoretic metrics for examining the Web site of a tourism destination with a focus on collaboration and cooperation among destination stakeholders. Paper [22] uses complex networks and chaos theory for analyzing two tourism destinations.

4 System Design

4.1 Data Set Preparation

Tourism is a highly informational industry where stakeholders are engaged in a collaborative business environment as prosumers of information about a large variety of tourism entities. In this paper we propose a prototype system that aims to analyze the implicit social network determined by the user's opinions, questions and answers that they express to describe their specific tourist experiences.

Our system uses a real data set that was extracted from AmFostAcolo[2] - a Romanian Web site having a similar purpose as IveBeenThere[3]. AmFostAcolo provides a large semi-structured database with post-visit tourist reviews about a large variety of tourist destinations covering specific aspects of accommodation, as well as general impressions about geographical places. The tourist information is organized as a tree-structured index according to the destination, region, section and place. Most often a destination represents a country, for example Romania. A region of Romania is for example Oltenia. A section can represent a locality (for example: the village Runcu) or a container of related places (for example "MĂNĂSTIRI din Oltenia" (En. "Monasteries of Oltenia")). A place can represent an accommodation unit or general impressions about a certain area (for example: "Sohodol Gorges").

Users are able to interact on AmFostAcolo in at least two ways. Firstly, users can ask questions and give answers related to a particular tourist place. This facility is useful whenever users would like to ask something not necessarily related to a specific impression. For example, a user can ask a question related to the tourist place of "Cheile Sohodolului". Secondly, users can post tourist impressions related to a certain place, so each place becomes a container of impressions or reviews written by the users registered at AmFostAcolo that visited the place and shared their impressions on the Web site. Each post can trigger discussions about the place in question. Other users can post echoes to this article. Most often these echoes are questions requiring clarifications or more details about the post. Questions in turn can trigger answers and so on.

Each user has associated a rank score that is calculated based on his or her activity portfolio and feedback received on the site, including: writing impressions, uploading photos, posting echoes to impressions, answering to questions, a.o.

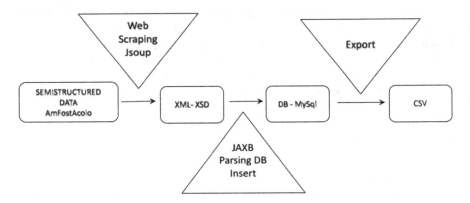

Fig. 1. Data extraction and preprocessing workflow

2 http://amfostacolo.ro/.
3 http://www.ive-been-there.com/.

In this paper we consider only the first situation, by focusing on the reconstruction of the social network based on the questions asked by users and answers given by users to these questions, related to each tourist place that is recorded in AmFostAcolo Web site. The second case, that considers also the echoes to impressions and the answers to echoes, was left as future work.

The task of data extraction was achieved using the data preprocessing workflow described in Fig. 1. This task contains the following activities:

i.   Data extraction from AmFostAcolo and its conversion to XML. This was achieved by developing a customized tool based on *jsoup* library for HTML processing[4].
ii. Data storage into a relational database developed using *mysql*[5]. This was achieved by developing a customized tool based on *jaxb*[6] for binding XML data with their Java implementation.
iii. Data export to CSV format, to facilitate its further processing using other tools.

4.2   Data Conceptual Schema

In this work we were interested in AmFostAcolo data related to users, their questions related to a tourist place, as well as the answers to these questions. For this purpose we have defined an XML schema for representing this data in XML. Data was extracted from AmFostAcolo using the technique of *Web scraping*, and then was saved onto a set of target XML files.

The XML schema of extracted data is presented in Fig. 2. The root note represents a *user*, so there is one such file for each user registered with AmFostAcolo. For each user we record the set of questions posted by this user, as well as their answers, using elements *qapost*, containing one *question*, as well as one *answer* element. For each answer we must also record the user that posted this answer, captured using the *qauthor* element. Using this simple representation we are able to capture the interactions established between the site users.

Using the resulting XML files, we were able to populate a *mysql* database with the relational schema presented in Fig. 3. This database can be queried directly using SQL for obtaining basic results about the data set. Furthermore, this database was exported into three CSV files *users.txt*, *question.txt* and *answer.txt*, that can be further used by other tools for graph-based and network analysis.

Currently our data set contains 54352 users, 53172 questions, and 53150 answers. The total size of the XML files is approximately 157 MB, while the total size of the CSV files is 67.3 MB. However, only a relatively small part of these users are involved in the question-answering process. We have counted 3132 users that asked questions, 7831 users that gave answers, and 8936 users that either asked questions, gave answers or both. So, the average number of questions per user, counting only users that gave questions, is 16.1, while the average number of answers per user, counting only the users that gave answers, is 6.7. The remaining users were involved in posting impressions, as

---

[4] http://jsoup.org/.
[5] https://www.mysql.com/.
[6] https://jaxb.java.net/library.

well as in posting echoes to impressions and answers to echoes. However, the analysis of these interactions was left as future work.

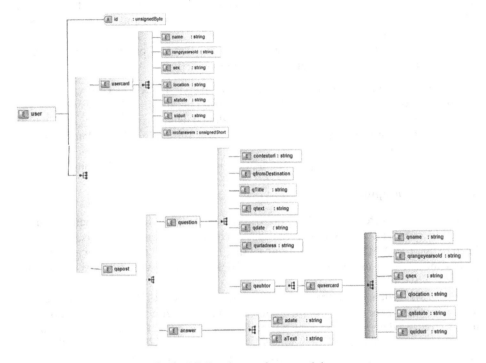

Fig. 2. XML schema of extracted data.

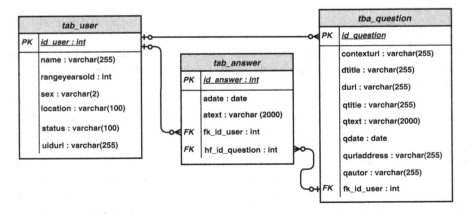

Fig. 3. Relational schema of extracted data.

4.3 System Block Diagram

Figure 4 illustrates the system block diagram that introduces the components and the processing steps of our system, including also the data preprocessing workflow. In this section we focus on the software interfaces with the other components of the system:

i. Data import into a network analysis tool, to enable network analysis of the data set.
ii. Data import into a graph database management system, to enable graph-based querying of the data set.

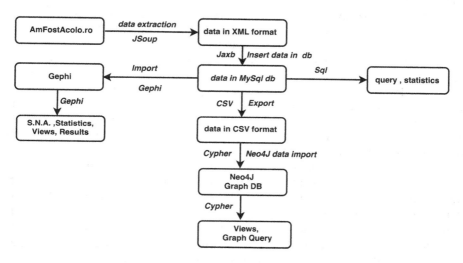

Fig. 4. System block diagram.

4.3.1 System Block Diagram

A new field of research, sometimes called Network Science (NS), was proposed for performing the analysis of complex interconnected systems, including social networks. The heart of this new research field leverages on methods from Graph Theory, Computer Science, and Statistics. Basically, NS identifies types of graphs that are suitable for modeling real-world phenomena, as well as metrics and computational methods to provide insight into the represented graph models. Network analysis tools are available for supporting the computational analysis of network-based models. One of our goals was to interface our system with such tools. We have chosen for exemplification the system *Gephi*[7] – a powerful graph analytics and visualization platform.

Our data set defines the implicit social network of AmFostAcolo Web site. Its nodes represent users, while its arcs represent interactions between users. The import of our data set into *Gephi* was achieved by using the *Gephi* function for importing data directly from a relational database server. Basically, we have created two temporary tables with nodes representing users and respectively the edges between those nodes, modeling the

---

[7] http://gephi.github.io/.

implicit interactions between users. Then, we have used the import function of *Gephi* by configuring the URL, the user, the password and the SQL queries needed to firstly retrieve the nodes, and secondly retrieve the edges, from the *mysql* relational database.

4.3.2   Data Import into a Graph Database

With the advent of social networks, linked data and semantic networks in the Web environment, there is a lot of research and industrial interest in new database models that better support the modeling of relationships between data, rather than the structure of the data [19].

Therefore, another goal of our work was to interface our system with a graph database tool, thus enabling a more natural graph-based querying of our data set. For that purpose we have chosen *neo4j*[8], one of the leading graph database systems.

A graph data model defines nodes and edges (or arcs) that connect nodes. Each node represents an entity, while each edge (or arc) represents a relationship between two entities. Entities (nodes) and relationships (represented as edges or arcs) have properties captured as key-value pairs, sometimes called attributes. A graph database supports data management based on the graph database model, by providing standard CRUD operations. One important advantage of graph databases is that they provide a natural support for querying interconnected nodes, without the need to define time consuming join operations.

We firstly defined a graph model of the AmfostAcolo social network. This model is illustrated in Fig. 5. It contains three entities represented by nodes *User*, *Answer*, and *Query*. Relationship *r:ASK* holds between the user *qUser* that asks the question and the *Question*. Relationship *r:HAS* holds between a *Question* and one of its answers *Answer*. Relationship *r:GIVEN_BY* holds between an *Answer* and the user *aUser* that provided the answer. Secondly, we have imported our data set expressed in CSV format into *neo4j*. The import step was achieved using the *neo4j* facility for importing data from tables represented in CSV format. The import was achieved in two steps. In the first step we have imported the nodes of the graph model: users, qustions, and answers, using one import statement for each type of node. For example, the statement for importing users is shown below:

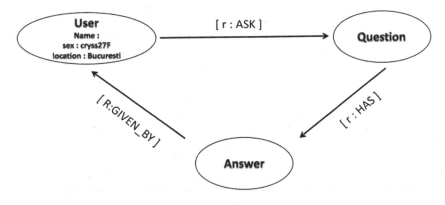

Fig. 5. *neo4j* data model.

---

[8] http://neo4j.com/.

```
USING PERIODIC COMMIT
LOAD CSV WITH HEADERS FROM "file:E:/csvafadb/tab_user.csv" AS row
CREATE (:User {userId: row.id_user, userName: row.name,
    userAge: row.rangeyearsold, sex: row.sex, location: row.location,
    status: row.status, uidurl:
```

In the second step we have imported the relationships, again using the "import from CSV" function of *neo4j*. For example, the statement for importing relationships between users and questions is shown below:

```
USING PERIODIC COMMIT
LOAD CSV WITH HEADERS FROM "file:E:/csvafadb/tab_question.csv" AS row
MATCH (user:User {userId: row.fk_id_user})
MATCH (question:Questions {qId: row.id_question})
MERGE (user)-[:ASK]->(question);
```

5 System Design

5.1 Querying the Graph Database

In this section we present an example for querying the *neo4j* graph database. For this purpose we have used Cypher [19]. The experiment was ran on *neo4j* version 2.2.2.

The first query determines the graph of users U that gave answers A to questions Q asked by a given user specified using his or her id. The query is shown below:

```
MATCH(uq:User)-[Ask]->(q:Questions)-[h:HAVE]->
    (a:Answers)-[r:GIVEN_BY]->(ua:USER]
    WHERE uq.userId="15"
    RETURN uq,q,a,ua
```

The result of this query is graphically illustrated in Fig. 6. The reference user is represented by the node in the center of the graph. It has 24 neighbors that represent the questions asked by this reference user. Additionally, these neighbors have 23 neighbors that are placed at depth two starting from the node that represents the reference user. These 23 neighbors represent the answers to the questions asked by the reference user. We can observe that the answers were given by 14 users as follows: 6 answers were given by 1 user, 4 answers were given by 1 user, 3 answers were given by 1 user, and each of the other 11 answers was given by another user.

The second query shows how we determined the friend-of-a-friend network of a given user. The query is shown below.

```
MATCH (uq:User)-[]->()-[]->()-[]->()-[]->()-[]->foaf
WHERE uq.userId="149"
RETURN foaf
```

If we would like to count the users of this network then instead of *RETURN foaf* we could use *RETURN count(foaf)*. In this particular case, the resulting network contains 586 users.

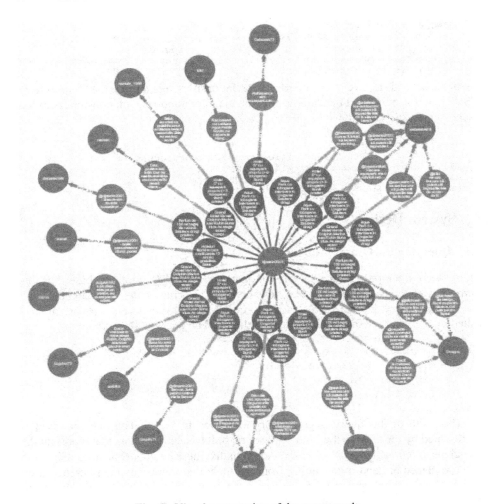

Fig. 6. Visual presentation of the query result.

5.2 Network Analysis Use Case

In order to understand the complex network of interactions that took place between the users of AmFostAcolo, we mapped the "question-answer" interactions to a directed graph. Intuitively a user is represented as a vertex. An arc from vertex A to vertex B is created when a question asked by user A gets a response from user B. For this experiment we only considered the users that participated in at least one "question-answer" interaction. Otherwise said, at least one of their raised questions received an answer or they gave at least an answer to a question raised by another user. Thus we obtained a graph with 8936 vertices and 29529 arcs that is presented in Fig. 7.

The analysis of this graph revealed 3 levels of granularity: entire graph, communities and vertices. At a first glance, we can consider the graph as a complex network of type

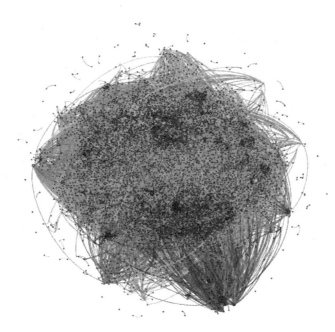

Fig. 7. AmFostAcolo "question-answer" user interaction network.

"core-periphery". It describes social non-trivial phenomena that cannot be explained only by graph theory. One phenomenon observed here is the "small-world", which means that the longest distance in the graph is relatively small (the diameter equals 14) compared to the total number of vertices (8936). A second phenomenon present in this graph is "rich get richer" (an explanation is given below).

We determined that this complex network is of type "core-periphery", as we could observe a "central well connected" community and some peripheral communities that are less connected. Analyzing the vertices that have a degree of 33 or more (419 vertices or 4.99 % of the total vertices), which is more than the average degree of 4.229, we can observe a significant increase in the graph connectivity from 0.011 to 0.064. Also the number of arcs between these vertices represents 13.81 % (4079) of the total number of vertices. These observations are consistent with the "core-periphery" type of complex network. This is a sign of a mature resilient community of users, i.e. a considerable amount of users would have to be removed in order the community to fall apart. Information usually travels very fast in this type of networks. It is known that negative information spreads even faster than positive information, thus a bad review can have a significant effect and reach in this community. Also, this type of network supports the more active users, i.e. the more a user gets involved (by asking questions and giving answers in this case), the more "important" he or she becomes.

When analyzing communities, we could observe that the top 3 communities contain almost 50 % of the users. The rest of nodes are scattered throughout the other 75 communities. The community detection was made with the help of the modularity algorithm [8] that is incorporated into *Gephi*. Further analysis must be done, but we suspect that these 3 communities have users that speak more general things or about more popular tourist places. A possible interpretation is that the majority of tourists share similar interests, while those with more specific interests are fewer. Note that tourists with more specific needs that are members of periphery communities can provide valuable and new information.

At the user or vertex granularity, we observed that the distribution of the degree is a power-law, see Fig. 8. For example, only 233 (~2 %) vertices have a degree of 50 or more and 6877 (~82 %) vertices have a degree between 0 and 10. This feature is determined by the "rich get richer" phenomenon, i.e. the more questions you ask the greater is the probability to get more answers. This phenomenon can be explained by the presence of a few users that answer a larger amount of questions compared to the rest. As Lada Adamic has proven [1], power-laws in the context of the World Wide Web should not surprise us as just a few web-sites have a large number of visitors. This distribution has also been found in many natural phenomena as Albert Barabási states [5].

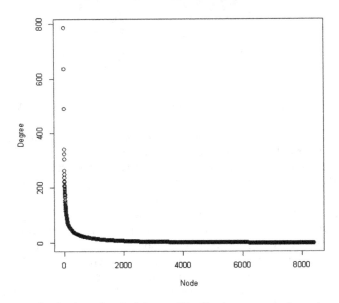

Fig. 8. Power law distribution of nodes' degree. The X axis represents the nodes and the Y axis represents the degree of each node.

6   Conclusions and Future Works

In this paper we reported our results about the construction and analysis of the implicit social network of a tourism information sharing Web site. The network is determined by the "question-answer" interactions between the users of the site. The data extracted from this Web site was firstly collected in XML format, then it was imported into graph database and network analysis tools. We present our initial conclusions regarding the type and characteristics of the resulted network. As future work we plan to expand our analysis in at least two directions: (i) to achieve a more complete quantitative analysis of the data set, by focusing on additional aspects, for example on community formation and network dynamics; (ii) to apply our techniques to other data sets extracted from the same or other tourism and non-tourism Web sites for content sharing.

References

1. Adamic, L.A., Huberman, B.A.: Power-law distribution of the world wide web. Science **287**(5461), 2115 (2000). doi:10.1126/science.287.5461.2115a
2. Al-Hassan, M., Lu, H., Lu, J.: A semantic enhanced hybrid recommendation approach: A case study of e-Government tourism service recommendation system. Decis. Support Syst. **72**, 97–109 (2015). doi:10.1016/j.dss.2015.02.001
3. Baggio, R.: The web graph of a tourism system. Physica A: Stat. Mech. Appl. **379**(2), 727–734 (2007). doi:10.1016/j.physa.2007.01.008
4. Bădică, C., Colhon, M., Şendre, A.: Sentiment analysis of tourist reviews: data preparation and preliminary results. In: Proceedings of the 10th International Conference on Linguistic Resources and Tools for Processing the Romanian Language, ConsILR 2014, pp. 135–142 (2014)
5. Barabási, A.-L.: Scale-free networks: a decade and beyond. Science **325**(5939), 412–413 (2009). doi:10.1126/science.1173299
6. Barabási, A.L., Gulbahce, N., Loscalzo, J.: Network medicine: a network-based approach to human disease. Nat. Rev. Genet. **12**(1), 56–68 (2011)
7. Becheru, A., Buşe, F., Colhon, M., Bădică, C.: Tourist review analytics using complex networks. In: Proc. of the 7th Balkan Conference in Informatics BCI 2015, Article 25. ACM (2015). doi:10.1145/2801081.2801088
8. Blondel, V.D., Guillaume, J.-L., Lambiotte, R., Lefebvre, E.: Fast unfolding of communities in large networks. J. Stat. Mech. Theory Exp. **2008**(10), P10008 (2008). doi:10.1088/1742-5468/2008/10/p10008
9. Borràs, J., Moreno, A., Valls, A.: Intelligent tourism recommender systems: A survey. Expert Syst. Appl. **41**(16), 7370–7389 (2014). doi:10.1016/j.eswa.2014.06.007
10. Büyüközkan, G., Ergün, B.: Intelligent system applications in electronic tourism. Expert Syst. Appl. **38**(6), 6586–6598 (2011). doi:10.1016/j.eswa.2010.11.080
11. Chung, N., Koo, C.: The use of social media in travel information search. Telematics Inf. **32**(2), 215–229 (2015). doi:10.1016/j.eswa.2010.11.080
12. Colhon, M., Bădică, C., Şendre, A.: Relating the opinion holder and the review accuracy in sentiment analysis of tourist reviews. In: Buchmann, R., Kifor, C.V., Yu, J. (eds.) KSEM 2014. LNCS, vol. 8793, pp. 246–257. Springer, Heidelberg (2014)
13. Erdős, P., Rényi, A.: On random graphs. Publicationes Mathematicae Drebrecen **6**, 290–297 (1959)
14. Garcia-Crespo, A., Lopez-Cuadrado, J.L., Colomo-Palacios, R., Gonzalez-Carrasco, I., Ruiz-Mezcua, B.: Sem-Fit: a semantic based expert system to provide recommendations in the tourism domain. Expert Syst. Appl. **38**(10), 13310–13319 (2011). doi:10.1016/j.eswa.2011.04.152
15. Gavalas, D., Konstantopoulos, C., Mastakas, K., Pantziou, G.: Mobile recommender systems in tourism. J. Netw. Comput. Appl. **39**, 319–333 (2014). doi:10.1016/j.jnca.2013.04.006
16. Granovetter, M.S.: The strength of weak ties. Am. J. Sociol. **78**, 1360–1380 (1973)
17. Gretzel, U., Werthner, H., Koo, C., Lamsfus, C.: Conceptual foundations for understanding smart tourism ecosystems. Comput. Hum. Behav. **50**, 558–563 (2015). doi:10.1016/j.chb.2015.03.043

18. Jiang, K., Yin, H., Wang, P., Yu, N.: Learning from contextual information of geo-tagged web photos to rank personalized tourism attractions. Neurocomputing **119**, 17–25 (2013). doi:10.1016/j.neucom.2012.02.049
19. McKnight, W.: Chapter Twelve – Graph databases: when relationships are the data. In: McKnight, W. (ed.) Information Management, pp. 120–131. Morgan Kaufmann, Boston (2014). doi:10.1016/B978-0-12-408056-0.00012-6
20. No, E., Kim, J.K.: Comparing the attributes of online tourism information sources. Comput. Hum. Behav. **50**, 564–575 (2015). doi:10.1016/j.chb.2015.02.063
21. Rodriguez-Sanchez, M.C., Martinez-Romo, J., Borromeo, S., Hernandez-Tamames, J.A.: GAT: Platform for automatic context-aware mobile services for m-tourism. Expert Syst. Appl. **40**(10), 4154–4163 (2013). doi:10.1016/j.eswa.2013.01.031
22. Scott, N., Baggio, R., Cooper, C.: Complex tourism networks. In: Scott, N., Baggio, R., Cooper, C. (eds.) Network Analysis and Tourism From Theory to Practice, Chap. 13, pp. 175–192 (2008)
23. Sigala, M., Chalkiti, K.: Investigating the exploitation of web 2.0 for knowledge management in the Greek tourism industry: an utilisation-importance analysis. Comput. Hum. Behav. **30**, 800–812 (2014). doi:10.1016/j.chb.2013.05.032
24. Wilson, C.: Searching for Saddam: a five-part serios on how the US military used social networking to capture the Iraqi dictator. Slate (2010)
25. Yang, W.-S., Hwang, S.-Y.: iTravel: a recommender system in mobile peer-to-peer environment. J. Syst. Softw. **86**(1), 12–20 (2013). doi:10.1016/j.jss.2012.06.041

A Mixed Approach in Recognising Geographical Entities in Texts

Dan Cristea[1,2], Daniela Gîfu[1(✉)], Ionuţ Pistol[1],
Daniel Sfîrnaciuc[1], and Mihai Niculiţă[3]

[1] Faculty of Computer Science, "Alexandru Ioan Cuza"
University of Iaşi, Iaşi, Romania
{dcristea,daniela.gifu,ipistol,
daniel.sfirnaciuc}@info.uaic.ro
[2] Institute for Computer Science Romanian Academy - The Iaşi Branch,
Iaşi, Romania
[3] Faculty of Geography, "Alexandru Ioan Cuza" University of Iaşi,
Iaşi, Romania
mihai.niculita@uaic.ro

Abstract. The paper describes an approach for automatic identification in Romanian texts of name entities belonging to the geographical domain. The research is part of a project (*MappingBooks*) aimed to link mentions of entities in an e-book with external information, as found in social media, Wikipedia, or web pages containing cultural or touristic information, in order to enhance the reader's experience. The described name entity recognizer mixes ontological information, as found in public resources, with handwritten symbolic rules. The outputs of the two component modules are compared and heuristics are used to take decisions in cases of conflict.

Keywords: Name entity recognition (NER) · Annotation conventions · Geonames · Geo-ontology · Gazetteer · Symbolic rules · Pattern matching techniques · Mixed approaches in NER

1 Introduction

MappingBooks is an on-going project[1] aiming to develop a new type of electronic product with a high impact in education and tourism. The main envisioned users are school pupils and students. The technology mixes methods from natural language processing, web cartography, web mapping, mixed reality techniques and ambient intelligence/ubiquitous computing to link mentions of geographical entities existing in school manuals onto data existing on the web, to localise these entities on 2D and 3D hypermaps [11] and to put them in correlation with the reader's location and related data. The toponyms can be supplemented with different type of information, diagrams or any related graphic materials. For example, if a reader is focusing a mention of the

---

[1] Financed by the Romanian Ministry of Education and Research (UEFISCDI) under the Partnerships Programme (PN II Parteneriate, competition PCCA 2013), project code: PN-II-PT-PCCA-2013-4-1878.

© Springer International Publishing Switzerland 2016
D. Trandabăţ and D. Gîfu (Eds.): EUROLAN 2015, CCIS 588, pp. 49–63, 2016.
DOI: 10.1007/978-3-319-32942-0_4

Mount Ceahlău in a school book, not only that a localisation of the Ceahlău Mountain will be signalled to her/him on an electronic map, but also information about this mountain, as a function of the context in which the toponym appears will be fetched and displayed on the user's mobile screen. If the school book covers the topic of physical geography, the localisation on a general map will be supplemented with thematic maps (geologic, landforms, climate maps), and associated thematic information.

What is most important in *MappingBooks* is that the connections from the book onto the virtual space would have to be realised by a technology, therefore automatically, and not by a human annotator. It is clear that this ambitious goal ought to be sustained by a powerful tool that manipulates with accuracy mentions of geographical entities in free texts. In this paper we describe that part of this project that deals with the recognition of name entities. Our approach mixes brute force methods (as provided by the use of a large collection of proper nouns) with symbolic methods (a collection of regular expressions, or rules, intended to discover the significance of proper names using the local context).

The paper has the following structure. In Sect. 2 we briefly present the state of the art in named entity recognition. Section 3 gives definitions for entities, as a semantic concept, and their realisation in texts. Section 4 shows the overall architecture of the system. Then, Sects. 5, 6 and 7 briefly describe the three component modules of the system, including some remarks about the evaluation. Section 8 describes the Name Entity Viewer, a Web service and application that facilitates visualisation of name entities in any text. Finally, Sect. 9 states some conclusions, while the Appendix gives a number of details about the other sources of free geographical data used in the project.

2 Background

The problem of extracting text information represents a constant concern in Natural Language Processing (NLP), now already for more than two decades. It has a wide range of applications in different domains (geography, biomedical sciences, business intelligence, etc.). The Message Understanding series of Conferences (MUC) has been launched at the beginning of 1990s to face the larger and larger interest of companies to extract, from unstructured text (such as newspaper articles), structured information about their activities or products. A few years later, at the 6th edition of MUC, the term *named entity* was introduced [8] and their recognition was considered a problem of classification. Named Entity Recognition (NER) and Classification became a task of Information Extraction. The important issue was the classification of words or word groups that signify proper names [18].

NER has become a very important topic for many other sub-fields of NLP [2, 15, 20]. Initially, the most important results were obtained using rule-based systems created manually. To overcome the tedious work of writing rules manually and to improve the rate of recognition, the researchers started to use statistical models, based on machine learning techniques, which have proved to be very effective. Here the manual effort was transferred in the direction of manual annotation, in order to build large corpora of positive examples. The most efficient techniques of our days, combine rule-based grammars with statistical (maximum entropy) models. An example of this type is the LTG system [17], presented at MUC-7. The FIJZ system [6], presented at the CONLL-2003 uses four different classifiers (robust linear classifier, maximum entropy, transformation-based learning, and hidden Markov model), which, combined under special conditions, produce very good results. Another notorious information extraction system was ANNIE (A Nearly-New Information Extraction System) [16], included in the GATE (General Architecture for Text Engineering) framework [5]. ANNIE recognizes person, location, organization, money, percent, data, address, identifier and unknown. ANNIE, was used with success for many languages, including Romanian, being a perfect example of combination of a lexical resource (gazetteer) and a rule-based approach in information extraction (a set of pattern/action rules written as JAPE grammars).

We mention, also, the MUSE system, incorporating ANNIE resources, processing also Romanian texts, as an example of a fast and cheap adaptation of an existing system to deal with new applications.

Another important project is TTL (Tokenizing, Tagging and Lemmatizing), a text processing platform developed at RACAI[2], trained to deal mainly with Romanian and English, which recognizes entities, does sentence splitting, tokenizing, chunking, etc. This platform works with techniques based on the use of regular expressions. In TTL, the NER function precedes the sentence splitter, avoiding thus the dangers of considering the dot in an abbreviation as signalling the limit of a sentence. Another NER system for Romanian that combines a collection of linguistic grammar rules and a set of resources is described in [10]. Other tasks are focused on: personal name disambiguation [14], named entity translation [7, 9] and acronym identification [19].

In the early 2000s, a priority in the research based on analysis of geographical references, focused on the named entity, was the geographic instances classification in text. For instance, the geographical references classification by assuming consecutive proper nouns as named entity candidates, using a co-training algorithm [3]. Also, a classification could be based on the fine-grained sub-types of geographical entities, knowing they refer generalized names as well as locations [23]. Some researchers suggested unsupervised learning methods in the NER area, related to bootstrapping learning algorithms [12, 13, 22]. Note that most bootstrapping approaches start with incomplete annotations and patterns derived from selected seeds, which imply possible annotation errors that can be included in the learning process. These errors could be avoided by designing statistical measures of control.

3   Entities in Text

In [4], we addressed the issue of annotating relations linking entities in texts. In the mentioned paper we say that any mention of an entity (restricted only to persons, gods, group of persons and of gods) is a mapping from a text expression to a corresponding

---

[2] The Romanian Academy Center for Artificial Intelligence, in Bucharest.

'container'[3]. In all corpus-based approaches, mentions, not containers, are annotated, but if semantic reasoning is tried on these annotations, then containers and their contents are recreated as semantic representations.

Following the usual tendency in the literature, we consider entities as being semantic categories expressed at the textual level by noun phrases (NPs). In certain contexts, proper names could be parts of NPs. It is therefore important to make the distinction between proper names and name entities. Proper names represent part-of-speech categories, therefore are manifested at the text level, while entities, among which also name entities, are semantic categories, therefore presented at a representational level. For instance, in the sequence *muntele Ceahlău* (*mount Ceahlău*), *Ceahlău* is a proper name, part of the NP *muntele Ceahlău*, but there is only one entity at the representational level, and this could be noted either [muntele Ceahlău] or [mount Ceahlău] (semantic representations are usually language independent). But NPs could have also a recursive structure, such that one NP may include one or more other NPs. In situations of the type $_{NP2}[\cdots_{NP1}[\cdots]_{NP1}\cdots]_{NP2}$ (in our examples an entity is noted as a span of text in-between square brackets and marked with a double label to ease reading: $_{NP1}[\text{span}]_{NP1}$ and heads are underlined words), NP1 and NP2 will both be marked if and only if head(NP1) $\neq$ head(NP2), as here: $_{NP3}[\text{clădirea}\ _{NP2}[\text{Universității}\ \ \text{din}\ \ _{NP1}[\underline{\text{Iași}}]_{NP1}]_{NP2}]_{NP3}$ ($_{NP3}[\text{the}$ $\underline{\text{building}}$ of $_{NP2}[\text{the}\ \underline{\text{University}}$ of $_{NP1}[\underline{\text{Iași}}]_{NP1}]_{NP2}]_{NP3}$).. Also, when talking about "imbricated entities" we will mean entities realised in text by imbricated (or nested) NPs. In the above example, two of the three entities are of a geographical semantic nature (GE)s: NP1 (a city) and NP2 (an organisation).

A discussion may arise in the case of complex expressions such as *Western and Central Europe*, which could be seen as a group entity [Western and Central Europe] or the juxtaposition of two simple entities, [Western Europe] and [Central Europe]. In order to avoid the proliferation of group entities, by combining in all possible ways the elements of similar geographical sets, we adopted the second solution, by disambiguating *Western* in the context of *Europe*, even if the component parts are separated by other tokens.

4   Approach and Architecture

A geographical entity is defined in our approach as a concept which can be associated with geographical characteristics, usually coordinates (point or bounding box) that are able to place it on the map, but potentially also: height, surface, population and others (see the Appendix for a comprehensive list of sources of additional information). In the context of our work, we look for geographical entities as referenced in texts, each textual reference being associated with specific geographical characteristics. Thus, from the perspective of our system, a text reference is equivalent to a geographical entity (which can have multiple equivalent text references). For all geographical entities

---

[3] A box or a container is associated with each character (entity), which in a text is, at the first mention, partially filled in with pieces of information and, subsequently, complemented with details (name, sex, kinship connections, composition, beliefs, religion, etc.).

annotated, we specify a type (general classification) and a subtype, specifying variations within a general type. In order to identify a geographical entity and its type and subtype, a three-step approach is used. The three steps are performed sequentially, as follows:

- a pre-processing phase performed over the original target document;
- a parallel application of a gazetteer module and a pattern-matching module;
- a merging and validation phase.

In *MappingBooks*, the pre-processing phase (PRE) involves several steps. First, the initial text is extracted from the original document (usually a PDF file including images and other non-textual content). This step involves the application of the iText[4] package, which leaves behind the text without formatting. Further on, the text is prepared by correcting diacritics and special characters, and eliminating end-of-line separators and other remains from the original format. Then, the corrected text is used as input for a chain of linguistic processes, adding the following markings: borders of lexical tokens, noun phrases and sentences, and part-of-speech categories and lemmas attached to tokens and compounds. For linguistic markings, the NLP-Group@UAIC-FII web-services[5] are used. The resulted annotated document (in stand-off XML format) serves as input for the next step, which passes through three other modules.

The gazetteer-applier (GAZ-APP) uses lists of toponyms and other geographic names, grouped by categories (usually called gazetteers – GAZ), to identify potential entity candidates. The result of this process is a document containing annotations for those surface names which are mentioned in GAZ, and where the type, subtype, coordinates, and other related geographical data, as found in the external resource, are added. Where ambiguous, a name will contain multiple tags, one for each category/subcategory and the disambiguation process is postponed.

In parallel, the patterns-applier module (PAT-APP) uses a set of patterns, described in terms of the markings left in the document by the PRE module, to discover potential geographical entities. The difference between PAT-APP and GAZ-APP is that the gazetteer makes use of strictly proper names, while the patterns include also contextual words that appear in their vicinity and which are used to reduce the ambiguities.

Finally, the merging and validation module (MER) compares the two annotated files to take final decisions of all markings.

5   The Gazetteer

We have looked for a gazetteer that includes as many Romanian names as possible. In a first step, we identified a set of types and their subtypes, and then we attached to them lists of relevant names. Our list includes 15 major types:

---

[4] http://itextpdf.com/.

[5] http://nlptools.info.uaic.ro/.

There are nine types of relationships between two tags:

1. LOCATION (with 23 subtypes, covering all locations that are usually referenced on the map of a region: cities, ports, streets, etc.);
2. GEO_POSITION (with 6 subtypes, corresponding to map references: parallel, meridian, cardinal point);
3. GEOLOGY (with 6 subtypes, indicating geological formations visible on a specific map);
4. LANDFORM (with 16 subtypes, covering types of physiographic formations usually indicated on maps: mountain, valley, cave, etc.);
5. CLIME (with 5 subtypes, covering meteorological data shown on some types of maps);
6. WATER (with 11 subtypes for each variation of surface aquatic formation: river, lake, strait, etc.);
7. DIMENSION (with 9 subtypes, corresponding to the various ways in which geographical entities can be accompanied by (exact or approximated) values in text: height, depth, surface, etc.);
8. PERSON (names of people, accompanied by professions, where specified);
9. ORGANISATION (with 5 subtypes: military, education, etc., indicating also possible locations associated with a particular organisation type);
10. URL (web references);
11. TIMEX (dates, moments of time, intervals, etc.);
12. RESOURCE (with 4 subtypes, for natural resources associated with locations);
13. INDUSTRY (with 4 subtypes, for industrial areas: factories, electrical plants, etc.);
14. CULTURAL (with 6 subtypes, for cultural areas: museums, parks, etc.);
15. UNKNOWN (for other geographical entities not covered by the above types).

In total, we identified 103 types + subtypes, out of which 67 are of a geographical nature. To populate our gazetteer organized around the above types, we consulted a number of freely available resources. Among them, Geonames[6] is commonly used by many developers who need accurate geographical reference data. Developed on the base of various governmental and educational data sources and completed with user contributed and verified data, this open resource provides now gazetteer data for over 2.8 million entities, with 5.5 million alternative names. For Romania, the focus of our developments, Geonames includes 25.951 names, with over 45.000 alternative names, with a density of ~ 0.108 toponyms/km^2, and 1 toponym to $\sim 1,000$ inhabitants. The names are grouped in 9 types, with 654 subtypes. The 9 types are identified by letters:

− A: country, state, region, …
− H: stream, lake, …
− L: parks, area, …
− P: city, village, …
− R: road, railroad, …

---

[6] http://www.geonames.org/.

- S: spot, building, farm, ...
- T: mountain, hill, rock, ...
- U: undersea, ...
- V: forest, heath, ...

For each of these types, besides the geographical coordinates, Geonames offers values for specific attributes, such as population (for P), surface (for A, H, L), height (for T), depth (for U), etc. In order to use these data to populate our gazetteer, we mapped our types/subtypes to those used by Geonames. As such, we found a many-to-one mapping between the 652 subtypes in Geonames and the 67 types/subtypes referring to the geography domain in our categorisation, to which are added the ones outside the domain of geography (DIMENSION, PEOPLE, ORGA-NISATION, URL, etc.). For example, any of the following type.subtype in Geonames is categorised as our type.subtype LANDFORM.HILL:

- T.BUTE - butte(s): a small, isolated, usually flat-topped hill with steep sides;
- T.HLL - hill: a rounded elevation of limited extent rising above the surrounding land with local relief of less than 300 m;
- T.HMCK - hammock(s): a patch of ground, distinct from and slightly above the surrounding plain or wetland;
- T.MND - mound(s): a low, isolated, rounded hill;
- T.PROM - promontory(-ies): a bluff or prominent hill overlooking or projecting into a lowland;
- T.MRN - moraine: a mound, ridge, or other accumulation of glacial till;
- U.HLLU - under-see hill: an elevation rising generally less than 500 meters.

The reduced number of types and subtypes in our classification theoretically should improve the precision of the GAZ-APP module, because of a lower classification ambiguity for each potential entity.

6   The Pattern-Matching Module

The set of patterns (PAT) of the PAT-APP module were manually written using the Graphical Grammar Studio (GGS)[7] tool [21]. GGS is a framework for the development and processing of grammars, which has incorporated a constraint description language allowing the implementation of composite features, of look-ahead and look-behind assertions, and placing priority scores on arcs, forcing thus a preference order in processing paths. GGS has been designed with the main purpose to perform syntactical and sub-syntactical analysis. Its networks consume and annotate sequences of tokens or other XML elements. The input tokens can include any number of associated attributes (usually denoting part of speeches, lemmas, articles in cases of nouns and adjectives, tokens IDs, etc.), which are mentioned in the GGS networks to specify acceptance conditions over the sequences they receive in input.

---

[7] http://sourceforge.net/projects/ggs/.

GGS networks are structured as directed graphs. The nodes of these graphs express token consuming conditions and are linked by directed edges. Some nodes can make jumps to other sub-graphs. The networks are meant to be integrated into NLP chains, since they usually require some sort of pre-processed input (tokens annotated in some form). A GGS network is basically a finite state machine whose nodes can be associated with states. The PAT-APP module is a matching process that takes as input a sequence of XML elements and a GGS network and tries to find a path in the network from its starting node to its ending node.

An example of how such a pattern can be viewed in GGS is shown in Fig. 1. The sequence *Bărăganul este cea mai mică câmpie* (EN: *Bărăganul is the smallest plain*) is parsed by the above pattern following path 5, resulting in the first word *Bărăganul* as being annotated as ENTITY with TYPE="LANDFORM" and SUBTYPE="PLAIN". Path 3 would match expressions like *câmpia cea mai mică este Bărăganul* (EN: *The smallest field is Bărăganul*), with the same annotations being added for *Bărăganul*.

The graphs are organized according to predetermined hierarchy of types, representing the 15 major categories, each of these presupposing the existence of other derivatives, with a total of 93 subcategories. A rule acts simultaneously for the identification and classification, combining contextual features found in tokens (like lemma, flexed word, etc.).

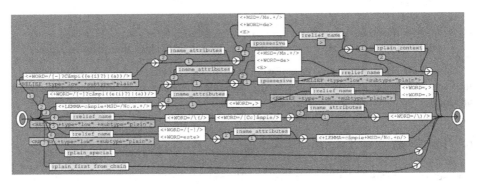

Fig. 1. A GGS network intended to recognize text references for geographical entities.

A concrete example when the GGS priority rules are applied is in the case of the sequence: *Universitatea Alexandru Ioan Cuza*. When processing this input, the pattern for the personal name recognition (*Alexandru Ioan Cuza* is a historic personage) has the lowest priority, and the preceding word *Univesitatea* forces the grammar to prefer a solution in which the whole expression is considered an educational institution: TYPE="ORGANISATION", SUBTYPE="EDUCATION".

Enumerations are treated for each of the types/subtypes considered. An example matching this rule is the sequence (see Fig. 2.): *oceanele, de la mare la mic, sunt după cum urmează: Pacific, Atlantic, Indian, Antarctic şi Arctic* (*the oceans, from biggest to smallest, are as follows:...*). In this case, every ocean is annotated as an ENTITY with TYPE="WATER" and SUBTYPE="OCEAN".

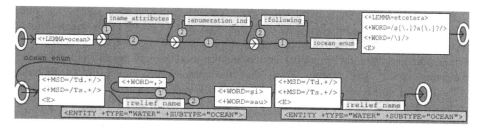

Fig. 2. A graph used to recognize an ocean name or an enumeration of such names, with the corresponding sub-graph for oceans that match elements of a list.

7 Merging, Validation and Evaluation

Table 1 shows a comparison between the GAZ-APP and the GGS-APP with respect to the number of occurrences they are able to recognise in the same text (only the main types are counted).

Table 1. Comparison of recalls for the GSS-APP and the GAZ-APP modules

Types	GGS	GAZ
câmpie (field)	47	56
chei (canyon)	2	–
continent (continent)	31	–
deal (hill)	31	–
deltă (delta)	29	58
depresiune (depression)	58	–
judeţ (county)	51	–
lac (lake)	34	28
luncă (meadow)	1	–
mare (see)	57	–
munte (mountain)	294	74
ocean (ocean)	2	–
oraş (city)	344	756
persoană_f_x_m (person)	52	–
persoană_feminin (person_fem)	8	–
persoană_masculin (person_masc)	35	–
podiş (plateau)	41	4
râu (river)	360	128
regiune (region)	114	220
sat (village)	240	934
ţară (country)	129	–
vârf_montan (peak)	21	118
Total	1981	2376

As can be seen, in general GGS-APP covers better certain categories than GAZ-APP, although, globally, GAZ-APP supersedes GGS-APP. This means that a proper treatment would be to combine the two processes. This observation actually let to the decision to include a merging and validation module, which follows both in the processing chain. The outputs from the two parallel processes, GAZ-APP and PAT-APP, are compared and validated by the MER module. The following cases are examined by MER:

- both GAZ-APP and PAT-APP annotate the same text span and the tag left by PAT-APP is among those left by GAZ-APP ⇒ the common tag is copied in the output file;
- both GAZ-APP and PAT-APP annotate the same text span and the tag left by PAT-APP is not among those left by GAZ-APP ⇒ the PAT-APP tag is copied in the output file;
- the text span annotated by GAZ-APP is included in the one annotated by PAT-APP and the tag left by PAT-APP is among those left by GAZ-APP ⇒ the common tag is copied on the largest text span in the output file;
- the text span annotated by GAZ-APP is included in the one annotated by PAT-APP and the tag left by PAT-APP is not among those left by GAZ-APP ⇒ the PAT-APP tag is copied on the largest text span in the output file;
- these is an intersection between the text spans annotated by the two modules and the tag left by PAT-APP is among those left by GAZ-APP ⇒ the common tag is copied on the union of the text spans in the output file;
- these is an intersection between the text spans annotated by the two modules and the tag left by PAT-APP is not among those left by GAZ-APP ⇒ the PAT-APP tag is copied on the union of the text spans in the output file;
- only one of the two modules annotate a certain text span with one or more tags ⇒ one tag out of those annotated is chosen randomly for that text span in the output file.

The criteria above show that, generally, more credibility is given to the PAT-APP module than to the GAZ-APP module, on the base that it uses the context to disambiguate names.

8   Web Applications and Services

In order to make an easy and better analysis of the system, a Name Entity Viewer was developed as a web application. The viewer is hosted on the *MappingBooks* project page[8]. It presents entities by highlighting them in different colors, as seen in Fig. 3.

Usually an annotator uses the entity viewer in tandem with a Name Entity Editor, also hosted by the *MappingBooks* project pages[9]. The user has the possibility to upload a new text file, let the system identify the entities and then correct them, by changing entity boundaries (continuous strings of tokens), their types and subtypes.

---

[8] http://85.122.23.18:8181/MappingBooks/resources/recognizer.
[9] http://85.122.23.18:8181/MappingBooks/resources/editor.

The default downloading format is *stand-off*. For example, for the text *Relieful României este definit de mai multe caracteristici* (*Romania's relief is defined by several characteristics*), the resulted XML is as follows:

```
<DOCUMENT>
<P ID="p1" offsetStart="0" offsetStop="58"/>
<S ID="s1" offsetStart="0" offsetStop="58"/>
<W Case="direct" Definiteness="yes" Gender="masculine" ID="w1.1"
 LEMMA="relief" MSD="Ncmsry" Number="singular" POS="NOUN" Type="
common" offsetStart="0" offsetStop="8" text="Relieful"/>
<W Case="oblique" Definiteness="yes" Gender="feminine" ID="w1.2"
 LEMMA="românie" MSD="Ncfsoy" Number="singular" POS="NOUN" Type=
"common" offsetStart="9" offsetStop="17" text="României"/>
<W EXTRA="intranzitiv" ID="w1.3" LEMMA="fi" MSD="Vmip3s" Mood="i
ndicative" Number="singular" POS="VERB" Person="third" Tense="pr
esent" Type="predicative" offsetStart="18" offsetStop="22"text="
este"/>
<W Case="direct" Definiteness="no" EXTRA="ParticipleLemma:defini
(tranzitiv)" Gender="masculine" ID="w1.4" LEMMA="definit" MSD="A
fpmsrn" Number="singular" POS="ADJECTIVE" offsetStart="23"offset
Stop="30" text="definit"/>
<W ID="w1.5" LEMMA="de" MSD="Sp" POS="ADPOSITION" offsetStart="3
1" offsetStop="33" text="de"/>
<W ID="w1.6" LEMMA="mai" MSD="Rg" POS="ADVERB" offsetStart="34"
offsetStop="37" text="mai"/>
<W Case="direct" Gender="feminine" ID="w1.7" LEMMA="mult" MSD="D
i3fpr" Number="plural" POS="DETERMINER" Person="third" Type="ind
efinite" offsetStart="38" offsetStop="43" text="multe"/>
<W Case="direct" Definiteness="no" Gender="feminine" ID="w1.8" L
EMMA="caracteristică" MSD="Ncfprn" Number="plural" POS="NOUN" Ty
pe="common" offsetStart="44" offsetStop="58"text="caracteristici
"/>
<ENTITY ID="e0" SUBTYPE="VILLAGE" TYPE="LOCATION" WORDSID="w1.2"
 offsetStart="9" offsetStop="17"/>
</DOCUMENT>
```

Developed as a web application, the Entity Editor is also platform-independent (available for a variety of operating systems including Windows, Mac OS and Linux).

An API allows the user to process books or large pieces of text and upload them on the site for subsequent queries. This implementation allows "live" entity type classifications, initiated by queries directly addressed by users[10]. The presented approach was adopted to make it suitable for online querying of huge texts.

---

[10] The API allows also identification of relations between entities, a facility not described in this paper.

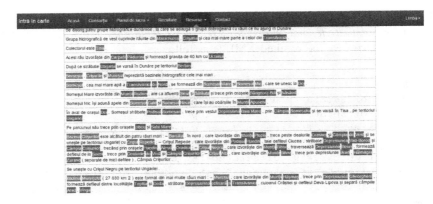

Fig. 3. The interface of the Entity Viewer

The web services are integrated in the MultiDPS platform [1], which is a Service-Oriented-Architecture that provides also tools for visualization of annotations, in a user-friendly manner.

9   Conclusions

We have presented in this paper an approach to build a sophisticated NER module for geographical entities that appear in Romanian texts. Its design is based on a combination between a brute-force approach (the use of an extended list of proper names) and a regular expressions approach (the use of a collection of manually written rules). The final decision to accept or reject an annotation over a span of words as being a geographical entity depends on the acceptance of more constraints, which are verified by a merge and validation module. For evaluation, the output of the merge module is compared against a test corpus, manually annotated. The results of the comparison are used to raise the quality of both resources (the gazetteer and the collection of patterns), in a bootstrapping enhancement loop, which is still on-going.

The work reported at this point is still preliminary and we don't want to risk conclusions regarding the accuracy of our system. However, the whole architecture is built on the presumption that the NER module could be made perfectible within the constraints imposed by the *MappingBooks* application.

We have a number of ideas that could guide an enhancement process: first, the more credibility that we give now to the PAT-APP module in its competition with the GAZ-APP module should be better contextualised and parameterised, by using more examples and training. Then, the random decision that we take now when we are left with more solutions should also be replaced by a biased decision, based on a thorough statistical study. Furthermore, the borders established for each entity should correspond to one of the NPs borders, i.e. the span of a name entity should always be equal to one existent noun phrase (conclusion left after examining the manually annotated corpus). But the NP-chunker is itself prone to errors and we believe that the result of the NER could correct the decisions previously taken by this module. Apparently, this kind of corrections should also be left at the handle of the MER module. When this study will be finished we hope to provide a thorough individual evaluation of the three component modules in relation with the manually annotated corpus.

Acknowledgement. The work was published with the support of the PN-II-PT-PCCA-2013-4-1878 Partnership PCCA 2013 grant *MappingBooks – Jump in the Book!*, having as partners the "Alexandru Ioan Cuza" University of Iaşi, SIVECO S.R.L. Bucharest and "Ştefan cel Mare" University of Suceava.

Appendix

It is worth mentioning that, in *MappingBooks*, the identified geographical entities are intended to be used as location points on the document, linking them with actual maps or external web links, or participating in relevant semantic relations. As a repository of spatial data GeoNetwork[11] was used, an open source platform that allows creating catalogues of spatial data, searching and storing their spatial metadata. The application is based on the principles of FOSS (Free and Open Source Software) and implements international standards (ISO/TC211 and OGC). The GeoNetwork application, running as a service server, stores the data in a database and provides a web interface through which the user can access catalogues of view spatial data and publishing spatial data, or can enter, visualise and edit metadata associated with the geospatial data.

Our intention is to attach to the recognised geographical entities different types of information, found on public sources. For this we are spotting a number of possible sources of free geospatial data: *Natural Earth*[12] – a set of cultural, physical and raster layers data, generalized for three spatial scales: 1:10 millions, 1:50 millions and 1:110 millions; *Romanian geomorphological regionalization*[13], digitised after a number of analogic versions; *Open Street Map*[14] – a dataset created by the community, open to anyone for contribution and editing, containing points of interests (POI), lines and polygons representing different types of spatial entities complemented with more information; *Bing Maps®*[15] – a product of Microsoft®, providing a WMS service with maps and aerial images and a Geocoding service, with a suite of data licenses, which, to some extent, can be used for personal and educational purposes; *Wikipedia*[16] containing in addition to the related text for each word, a location, as geographical coordinates, for toponyms; the Romanian SDI and the Romanian INSPIRE geoportal[17] crated by the *National Agency for Cadastre and Registration*[18], through the National Geodetic Fund

---

[11] http://geonetwork-opensource.org/.

[12] http://www.naturalearthdata.com/.

[13] http://earth.unibuc.ro/download/harta-unitati-relief-romania.

[14] http://www.openstreetmap.org.

[15] www.bing.com/maps.

[16] http://ro.wikipedia.org.

[17] http://geoportal.ancpi.ro.

[18] ANCPI – http://www.ancpi.ro.

and several collaborations; *Data.gov.ro* – a portal of partially geospatial data produced by Romanian government agencies (the SIRUTA national codes for administrative units); statistical data provided by the *National Statistics Institute*[19] – the Romanian national statistics service, linkable to geospatial boundaries: the TEMPO database[20], the eDemos database[21], the IDDT database[22] of sustainable development indexes[23], etc.

Also, *compiled datasets* can be produced by linking statistical databases with geospatial data, or through generalisation or other kinds of spatial analysis. For most datasets global processing is needed for cutting the region of interest, or to possibly change the format and projection.

References

1. Anechitei, D.A.: MultiDPS - a multilingual discourse processing system. In: Proceedings of the COLING 2014, the 25th International Conference on Computational Linguistics: System Demonstrations (COLING 2014), Dublin, Ireland, August 2014, pp. 44–47 (2014)
2. Borthwick, A., Sterling, J., Agichtein, E., Grishman, R.: Exploiting diverse knowledge sources via maximum entropy in named entity recognition. In: Proceedings of the 6th Workshop on Very Large Corpora (1998)
3. Collins, M., Singer, Y.: Unsupervised models for named entity classification. In: Proceedings of the Joint SIGDAT Conference on Empirical Methods in Natural Language Processing and Very Large Corpora (EMNLP/VLC), College Park, MD, pp. 100–110. Association for Computational Linguistics (1999)
4. Cristea, D., Gîfu, D., Colhon, M., Diac, P., Bibiri, A.-D., Mărănduc, C., Scutelnicu, A.L.: Quo vadis: a corpus of entities and relations. In: Gala, N., Rapp, R., Enguix, G.B. (eds.) Language Production, Cognition, and the Lexicon. Springer International Publishing, Cham (2015)
5. Cunningham, H., Maynard, D., Bontcheva, K., Tablan, V.: GATE: a framework and graphical development environment for robust NLP tools and applications. In: Proceedings of the 40th Anniversary Meeting of the Association for Computational Linguistics (2002)
6. Florian, R., Ittycheriah, A., Jing, H., Zhang, T.: Named entity recognition through classifier combination. In: Proceedings of the Seventh Conference on Natural language learning (HLT-NAACL 2003), vol. 4, pp. 168–171 (2003)
7. Fung, P.: A Pattern Matching Method for Finding Noun and Proper Noun Translations from Noisy Parallel Corpora. In: Proceedings of the Association for Computational Linguistics (1995)
8. Grishman, R., Sundheim, B.: Message understanding conference - 6: a brief history. In: Proceedings of the COLING (1996)

---

[19] http://www.insse.ro.

[20] https://statistici.insse.ro/shop/?lang=ro.

[21] http://edemos.insse.ro/portal.

[22] http://www.insse.ro/cms/files/IDDT%202012/index_IDDT.htm.

[23] http://www.insse.ro/cms/files/Web_IDD_BD_ro/index.htm.

9. Huang, F.: Multilingual named entity extraction and translation from text and speech. Ph.D. thesis, Carnegie Mellon University (2005)
10. Iftene, A., Trandabăţ, D., Toader, M., Corîci, M.: Named entity recognition for Romanian. In: Proceedings of the 3th Conference on Knowledge Engineering: Principles and Techniques Conference (KEPT 2011), pp. 19–24, vol. 2. Studia Universitatis, Babeş-Bolyai, Cluj-Napoca (2011)
11. Kraak, M.-J., Rico, V.D.: Principles of hypermaps. Comput. Geosci. **23**(4), 457–464 (1997)
12. Lee, S., Lee, G.G.: A bootstrapping approach for geographic named entity annotation. In: Myaeng, S.-H., Zhou, M., Wong, K.-F., Zhang, H.-J. (eds.) AIRS 2004. LNCS, vol. 3411, pp. 178–189. Springer, Heidelberg (2005)
13. Li, H., Srihari, R.K., Niu, C., Li, W.: InfoXtract location normalization: a hybrid approach to geographic references in information extraction. In: Proceedings of the HLT-NAACL 2003 Workshop on Analysis of Geographic References, Alberta, Canada, pp. 39–44 (2003)
14. Mann, Gideon S. and Yarowsky, D.: Unsupervised Personal Name Disambiguation. In: Proceedings of the 9th Conference on Computational Natural Language Learning (2003)
15. Masayuki, A., Matsumoto, Y.: Japanese: named entity extraction with redundant morphological analysis. In Proceedings of the Human Language Technology Conference – North American chapter of the Association for Computational Linguistic (2003)
16. Maynard, D., Tablan, V., Ursu, C., Cunningham, H., Wilks, Y.: Named entity recognition from diverse text types. In: Recent Advances in Natural Language Processing 2001 Conference, Tzigov Chark, Bulgaria, pp. 257–274 (2001)
17. Mikheev, M., Grover, C. and Moens, M.: Description of the LTG system used for MUC-7. In: Proceedings of the 7th Message Understanding Conference (MUC-7) (1998)
18. Nadeau, D., Sekine, S.: A survey of named entity recognition and classification. In: Sekine, S., Ranchhod, E. (eds.) Named Entities: Recognition, Classification and Use, vol. 30(1), pp. 3–26 (2007). Special issue of Lingvisticæ Investigationes
19. Nadeau, D., Turney, P.A.: Supervised learning approach to acronym identification. In: Proceedings of the 18th Canadian Conference on Artificial Intelligence (2005)
20. Sekine, S., Grishman, R., Shinnou, H.: a decision tree method for finding and classifying names in Japanese texts. In: Proceedings of the Sixth Workshop on Very Large Corpora (1998)
21. Simionescu, R.: Graphical grammar studio as a constraint grammar solution for part of speech tagger. In: Proceedings of the International Conference Resources and Tools for Romanian Language – ConsILR-2011, Bucharest. "Alexandru Ioan Cuza" University of Iaşi Publishing House (2011)
22. Smith, D.A., Mann, G.S.: Bootstrapping toponym classifiers. In: Proceedings of the HLT-NAACL 2003 Workshop on Analysis of Geographic References, Alberta, Canada, pp. 45–49 (2003)
23. Yangarber, R., Lin, W., Grishman, R.: Unsupervised Learning of Generalized Names. In: Proceedings of the 19th International Conference on Computational Linguistics (COLING 2002), Taipei, Taiwan, pp. 1135–1141 (2002)

User Profiling and Assessing the Suitability of Content from Social Media

Image and User Profile-Based
Recommendation System

Cristina Şerban, Lenuţa Alboaie, and Adrian Iftene$^{(\boxtimes)}$

Faculty of Computer Science, "Alexandru Ioan Cuza"
University of Iaşi, Iaşi, Romania
{cristina.serban,adria,adiftene}@info.uaic.ro

Abstract. A great variety of websites try to help users in finding items of interest by offering a list of recommendations. It has become a function of great importance, especially for online stores. This paper presents a recommendation system for images which works with ratings to compute similarities, and with social profiling to introduce diversity in the list of suggestions.

Keywords: Image similarity · Collaborative filtering · Social profiling · Recommendation dichotomy

1 Introduction

Recommendation systems represent a class of Web applications that predict user responses to options [1]. They automatically predict the information or items that may be of interest to a user and help in overcoming information overload by personalizing suggestions based on likes and dislikes. Such systems can be found in many online sites, especially online stores (e.g. Amazon, eBay, Barnes & Noble, IMDb, YouTube), making it much easier to explore the various available options and to find items of interest. They are valuable as they reduce the cognitive load on users and play a part in introducing quality control.

This paper describes an image recommendation system that uses similarity between items, similarity between users and social profiling to predict what other images a user might enjoy. In Sect. 2 we present an overview over recommendation systems. Section 3 presents our solution, covering the way we gathered the data and how the system creates recommendations. A use case is shown in Sect. 4, while Sect. 5 demonstrate how well our system works, by performing both automated validation and user validation. Finally, Sect. 6 lists our conclusions and possible future work.

2 Recommendation Systems Overview

In a recommendation system there are two types of entities: *users* and *items*. Users have preferences for certain items and it is these preferences that such a system must identify. The data available to the system is represented as a utility matrix [1]. It contains, for each user-item pair, a value that represents the degree of preference of the

© Springer International Publishing Switzerland 2016
D. Trandabăţ and D. Gîfu (Eds.): EUROLAN 2015, CCIS 588, pp. 67–82, 2016.
DOI: 10.1007/978-3-319-32942-0_5

user for the respective item. This matrix is generally sparse and the goal of a recommendation system is to predict the values in the blank entries. However, it is not always necessary to predict every such entry, but only those entries in each row that are likely to contain high values [1].

There are a number of different technologies used by recommendation systems, but two broad groups can be distinguished [1]. *Content-based* systems examine the features or properties of the suggested items. They recommend items that are similar in content to other items the user has previously expressed interest in. *Collaborative filtering* systems provide recommendations using various similarity measures between users and/or items. They collect human judgements in the form of ratings for items and exploit the similarities and differences of user profiles when selecting what to suggest [2]. The recommended items for a user are the ones preferred by other similar users.

In constructing our system, we approached the problem in a manner similar to [3], by combining item-based and user-based collaborative filtering and allowing each of these techniques to compensate when the other produces few or no results. We have also created recommendations based on social data, in order to encourage diversity and avoid the echo chamber effect.

3   The Proposed System

3.1   Gathering Data and Creating Image Profiles

The system used to gather the data allows a user to sign in using the account given by the faculty. There are 100 images with different (but not necessarily unrelated) themes available. The user is randomly given an image that they have not tagged yet (as shown in Fig. 1) and they must give a rating and add at least two tags. The scale for rating an image is:

- 5 - I like it very much;
- 4 - I like it;
- 3 - Neutral;
- 2 - I dislike it;
- 1 - I dislike it very much.

This system was given to a diverse group of students (both Bachelor and Master's) from the Faculty of Computer Science, "Alexandru Ioan Cuza" University of Iaşi. A total of 78 students tagged and rated all 100 images in our collection and we used this information to automatically assign a list of tags to each picture [4].

The lemmas from the tags were extracted using Stanford CoreNLP, a suite of natural language analysis tools created by The Stanford Natural Language Processing Group [5]. The stopwords were eliminated and, for each image, the frequency of all lemmas associated with it was computed. From this result, only those lemmas that appeared at least 5 times (i.e. at least 5 students associated that word with the images) were retained. This threshold helps with ignoring wrong or misspelled words, as it is highly unlikely that 5 people have wrongly spelled the same word in the same way, and also with eliminating outliers in terms of relevance (i.e. tags that are loosely related to the images). The result contained an average of 12,67 tags per image.

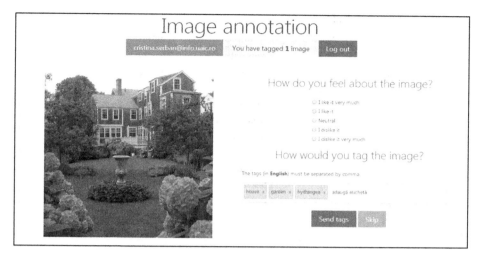

Fig. 1. A small portion of the images and tags graph, where the nodes containing numbers are images (the numbers being their id) and the nodes containing text are tags.

Furthermore, all the 100 images were also tagged by us. We decided that the order of the tags will dictate their relevance (i.e. the first is the most relevant) and we combined the two results by adding the extra tags from our lists to the ones obtained from the students, the position of insertion depending on how relevant we felt that these additional tags were. After this enrichment, there were, on average, 13.64 tags per image. Thus, for each image, there was, on average, approximately one tag added. This is an indication that the automatic selection of tags from the ones given by the students yielded a fairly good result, which needed very little adjustment.

3.2 Recommendations Based on Image Similarity

We created an undirected graph where the vertices are either images or tags. There are edges from images to tags and between tags. There are no edges between images. The edges represent different types of relationships between vertices and have weights (see Table 1).

An image may be connected to multiple tags and a tag may be connected to multiple images. We call this type of relationship an *Annotation* and its weight depends on the position of the tag in the list for the image and the total number of tags associated with that image. We call this the *relevance* of a tag for an image and it is computed as shown in (1), where *tag_count(i)* is the total number of tags for image i and *position(t, i)* is the position of tag t in the list of tags for image i.

$$relevance(t, i) = \frac{(tag\_count(i) - tag\_position(t, i) + 1)}{tag\_count(i)} \qquad (1)$$

Table 1. Types of edges and their weight

Node 1 (N_1)	Node 2 (N_2)	Relationship type	Weight
Image	Tag	Annotation	$relevance(N_2, N_1)$
Tag	Tag	Synonym	0.9
Tag	Tag	Subword	0.8
Tag	Tag	Attribute	0.7
Tag	Tag	Nominalization	0.7
Tag	Tag	Hypernym	0.6
Tag	Tag	Common words	0.4
Tag	Tag	Similar to	0.3
Tag	Tag	See also	0.2
Tag	Tag	Indirect synonym	0.15

There are nine types of relationships between two tags:

- *Subword* - A tag is a word contained inside another tag that is a phrase;
- *Common words* - The two tags are phrases that contain common words;
- *Attribute, Nominalization, Hypernym, Similar to, See also* - They were extracted using WordNet 3.0 [6] and have the same meaning as the respective pointer types [7] in WordNet. If tag t_1 and tag t_2 are in one of these relationships, it means that one of the synsets of t_1 has a pointer of this type to one of the synsets of t_2, or the other way around;
- *Synonym* - If tag t_1 and tag t_2 are in this relationship, it means that t_2 belongs to one of the synsets of t_1, or the other way around;
- *Indirect synonym* - If tag t_1 and tag t_2 are in this relationship, it means that their synsets have words in common.

Using this graph, it is possible to start in an image node and reach other image nodes by passing through several tag nodes. Along a path, the weights of the edges are multiplied and the value with which an image node is reached represents the similarity between the image from which we started and the image to which we arrived.

Let i_s be the start image node, i_e the end image node, and $t_1, t_2, ..., t_n$ the tags on the path from i_s to i_e. The similarity score between i_s and i_e is computed as shown in (2). If there is more than one path from i_s to i_e, the final similarity score will be the highest score among the paths.

$$similarity(i_s, i_e) = \max_{t_1,...,t_n}\left\{ relevance(t_1, i_s) \cdot \left(\prod_{i=2}^{n} weight(t_{i-1}, t_i)\right) \cdot relevance(t_n, i_s)\right\} \quad (2)$$

The graph, a small portion of which is depicted in Fig. 1, was stored using Neo4j, a highly scalable and robust native graph database [8]. In Fig. 2, the nodes containing numbers are images (the numbers being their id) and the nodes containing text are tags. The types of relationships between the tags can be easily observed.

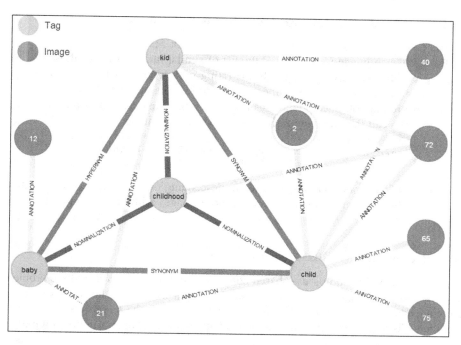

Fig. 2. A small portion of the images and tags graph, where the nodes containing numbers are images (the numbers being their id) and the nodes containing text are tags.

For example, let image 2 be our starting node and image 12 our end node. Image 2 has two tags, *child* and *kid*. From image 2 to image 12, there are several paths available, for instance:

- 2 → kid → baby → 12;
- 2 → child → baby → 12;
- 2 → child → childhood → baby → 12.

The final similarity score between images 2 and 12 will be the maximum value computed by multiplying the weights of the edges along each path. In order to find the path of maximum score, we used a slightly altered version of Dijkstra's shortest path algorithm. All nodes are initialized with a score of 0, except for the starting image node, which is assigned a score of 1. When going along a path, the score of the current node is multiplied with the weight of the edge and, if the result is greater than the score for the next node, it is updated. We do not move forward when we reach an image node. After the exploration is done, the image nodes with scores greater than 0 form the similarity list for the start image node.

For a user u the recommendation process works as follows:

- We take all images that were given a rating of 4 or 5;
- For each such image i_1, we look at the list of similar images, discard those that the user has already rated and multiply the rating of i_1 with the similarity score, thus obtaining a prediction for how the user would rate the new image i_2;
- If there are multiple such predictions for i_2, we retain the maximum value, as shown in (3). We call this the recommendation score for that image;

$$rec\_score(i_2, u) = \max_{i_1 \neq i_2, \, i_1 \text{ ratedby} u} \{similarity(i_1, i_2) \cdot rating(u, i_1)\} \qquad (3)$$

- The list of recommendations is sorted in descending order of the recommendation scores. We only take into consideration those images with scores higher than 3.5.

3.3   Recommendations Based on User Similarity

A popular method for doing user-based collaborative filtering is to regard the problem as a machine learning one and use a classifier such as k-NN. This means that the recommendations for a user will be an aggregation of what the k users most similar to them have liked. Systems that operate with a large number of users usually group them into clusters and create recommendations for each cluster and not each individual user. Furthermore, working with groups of items instead of individual ones may also be necessary when their number is big.

In our system, however, since the image collection is relatively small and hetero-geneous, and the amount of users is manageable, we decided to work with individual users and individual images. When computing the similarity between two users, we look at the images that they both have rated and compare the two vectors of ratings. For this purpose, we used Pearson's correlation coefficient (see [9], Chap. 7). Pearson correlation coefficient is a measure of the linear dependence between two variables X and Y and takes values between -1 and 1, inclusive. A value of -1 is total negative correlation, 0 is no correlation and 1 is total positive correlation.

In order to obtain good recommendations, we had to decide on a value for k. We used the ratings given by the 78 students mentioned before as training data and con-ducted a series of experiments to give us insight into how the system would behave for different values. We chose the following approach:

- For each user in our training set of 78, a random subset of their ratings was selected;
- The similarity between the current user and the other 77 was computed using both this subset (predicted similarity) and the entire set of ratings (real similarity);
- For various values of k, ranging from 1 to 15, we computed the mean of the squared errors between the predicted similarity value and the real similarity value for the first k results, in descending order;
- This process was repeated for different sizes of the random selection of ratings: 8, 10, 15, 20, and 25. Three different runs were created for each value and were combined into a single result by computing the average of the errors;

What we wanted to see was which values of k better predict the similarity between users when the number of known ratings varies. Our intuition was that, when there are few ratings given by the user, we should look at more of his neighbors than when there is a fairly moderate amount of already given ratings.

This was proven by the results obtained, as seen in Fig. 3. For each selection size, the squared error mean decreases as k increases, small values of the selection size having the steepest descent. For bigger selections, the errors obtained for all values of k are very close in value, while for smaller selections, they are not. Based on these results, we decided to use:

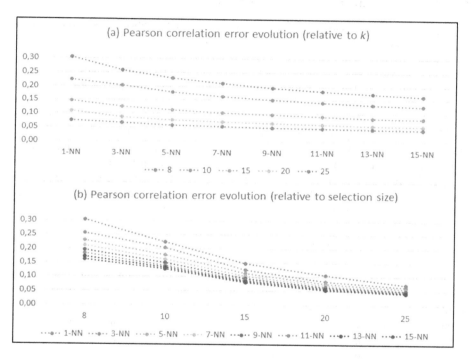

Fig. 3. Error evolution for Pearson correlation: (a) relative to k; (b) relative to the size of the ratings selection

- $k = 7$ for sizes smaller than 15 (because, for $k > 7$, the error varies very little), and
- $k = 5$ for sizes equal or greater than 15 (because, for $k > 5$, the error varies very little).

Let N be the set of k-nearest neighbors for a user u and N_i the subset of these neighbors that have given ratings for an image i. The recommendation score for i is computed as shown in (4).

$$rec\_score(i, u) = \frac{\sum_{n \in N_i} rating(n, i) \cdot similarity(u, n)}{\sum_{n \in N} similarity(u, n)} \qquad (4)$$

Note that if $N_i \subset N$, the formulas will act as if the neighbors in $N\backslash N_i$ have given a rating of 0. We want this behavior because we are more interested that those images rated by more neighbors be higher in the list because we are certain about whether they like them or not, as opposed to the situation when they have not expressed their interest in the image. The list of recommendations for user u (in descending order of the recommendation scores) comes from the k users most similar to u, for images not rated by u. We only take into consideration those images with scores higher than 4.

3.4   Recommendations Based on Social Similarity

The approaches previously described will yield recommendations based on what the user has already liked, regardless of whether they are found by exploring similar images or similar users. This could lead to an echo-chamber of what the user has explicitly expressed interest in and not predict some other kind of images that the user might also like, but they have not had the chance to show they are interested in yet.

In an attempt to avoid this, we decided to include in our recommendation list some images that are not necessarily related to those that the user has previously given a high rating to, but have the potential to be of interest to them. Thus, from the set of 78 users that we have used for training, 52 had Facebook accounts that we were able to access. We extracted their lists of interests and their relationships on the network.

The interest similarity between users was computed using (5). We chose this approach and not the Jaccard distance (see [1], Sect. 3.1.1) because we feel that the similarity of a to b is not the same as the similarity of b to a.

$$social\_similarity(a, b) = \frac{interest\_similarity(a, b)}{distance(a, b)} \qquad (5)$$

We also extracted the distance between users, i.e. the length of the shortest path from one user to another. The social similarity between two users is their interest similarity divided by their distance - see (6). However, since the formula we used for interest similarity is not symmetric, neither will the social similarity be. Thus, in general, $social\_similarity(a, b) \neq social\_similarity(b, a)$ – the case when a and b have the same interests being the exception.

$$interest\_similarity(a, b) = \frac{|interests_a \cap interests_b|}{|interests_a|} \qquad (6)$$

To obtain recommendations for a, we multiply the social similarity score of a and b with the ratings for images that b has liked (i.e. that were rated 4 or 5). We discard the images that a has already rated. If there is more than one score obtained for an image, we retain the maximum, as shown in (7).

$$rec\_score(i, a) = \max_{b \neq a, b\,rated\,i}\{social\_similarity(a, b) * rating(b, i)\} \qquad (7)$$

From our runs, we have observed that it is very likely that many images will have the same score. In case there are more images with the same score than the maximum number of recommendations the system has to select, then they are chosen randomly.

4 Recommendation Use Cases

For a user, the final output of our system is a list of 10 recommendations. It will try to select 4 based on image similarity, 3 based on user similarity and 3 based on social relationships. However, if there are not enough recommendations of one type, the system will select more from the others, if possible. We will look at how the results are obtained for two test users, to whom we will refer as user A and user B.

Table 2. Ratings given by user A

Rating	Count	Image list
5	7	1, 9, 12, 13, 16, 93, 100
4	4	17, 20, 24, 28
3	1	41
2	3	44, 68, 96
1	1	55

Fig. 4. Ten of the pictures that were given high ratings by user A.

Table 3. Use case A: results for image similarity-based recommendations

Score	Image list	Score	Image list	Score	Image list
5	10, 81	4,31	31	4,125	64
4,44	95	4,263	84	4,06	62
4,43	58	4,26	18	4	90

4.1   Use Case A: Combining All Three Types of Recommendations

Initially, user A gave the 16 ratings listed in Table 2. Images 1, 9, 12, 13, 16, 93, 100, 17, 20, and 24 are shown in Fig. 4.

For image similarity, the system obtains a total of 21 recommendations, the first 10 with the highest score being shown in Table 3.

Fig. 5. Maximum score path from image 9 to image 10

As an example, let us look at the score for image 10. The user has given a rating of 5 to image 9 and "butterfly" is the first tag for both 9 and 10. The path between the two images is depicted in Fig. 5. The similarity score between the two images is 1 and the recommendation score for image 10 will be $1 \times 5 = 5$.

For user similarity, the system obtains the 8 recommendations in Table 4. Let N_1, N_2, N_3, N_4 and N_5 be the 5-nearest neighbors of A. The Pearson correlation values for

Table 4. Use case A: results for user similarity-based recommendations

Score	Image list	Score	Image list
5	35	4,19	72
4,80	79	4,06	88
4,42	31	4,03	61
4,39	64	4,01	52

these neighbors are 0,889, 0,761, 0,741, 0,729, and 0,691, respectively. Image 35 obtained a score of 5 because all five neighbors gave it a rating of 5. Image 79 received a rating of 4 from N_3 and ratings of 5 from the other neighbors, thus its score is $(5 \times (0,889 + 0,761 + 0,729 + 0,691) + 4 \times 0,741)/(0,889 + 0,761 + 0,741 + 0,729 + 0,691) \approx 4,80$.

Fig. 6. Recommendations for user A

For recommendations based on social data, the system obtains a list of 84 (i.e. all the unrated images, with associated scores). The maximum score was 1,548 and there were 15 images that obtained it - 11, 21, 25, 33, 37, 45, 53, 64, 66, 69, 75, 80, 87, 91, and 99. This is because we did not impose a minimum score for this criterion, as we prefer to always take some from the top, if possible.

Since there are enough recommendations in every list, the system will select:

- 4 images from the image similarity list - 10, 81, 95, and 58;
- 3 images from the user similarity list - 35, 79, and 31;
- 3 images from the social similarity list - 75, 91 and 37 in this particular run, randomly chosen from those with the highest score.

Table 5. Feedback of user A on the recommendations given to them

Image	10	81	95	58	35	79	31	75	91	37
Feedback	5	5	5	5	5	5	5	3	5	4

This list of results (shown in Fig. 6) was given to user A for feedback rating. Their response is listed in Table 5. Note that the few lower ratings that user A gave were for

Table 6. Ratings given by user B

Rating	Count	Image list
5	4	29, 95, 98, 100
4	4	43, 53, 80, 86
3	7	6, 11, 15, 31, 50, 83, 84, 94
2	2	7, 38
1	1	74

Fig. 7. Ten pictures that were given high and neutral ratings by user B

images that were recommended based on social similarity, but those 3 selections were nevertheless good, while the other 7 were highly appreciated.

Table 7. Use case B: results for image similarity-based recommendations

Score	Image list	Score	Image list	Score	Image list
4,545	81	4	33, 12	3,78	82
4,44	16	3,93	79	3,73	61
4,074	64	3,88	1	3,53	19

4.2   Use Case B: Combining Two Types of Recommendations

User B initially gave the 19 ratings listed in Table 6. Images 29, 95, 98, 100, 43, 53, 80, 86, 6 and 11 are shown in Fig. 7.

Fig. 8. Maximum score path from image 95 to image 81

For image similarity, the system obtains a total of 10 recommendations, listed in Table 7. This low number of recommendations is justified by the fact that user B has given ratings of 4 and 5 to few images (i.e. 8 out of 19).

Image 81 has the score 4,545 because its second tag is "flower", which is the first tag for image 95. Because image 81 has 11 tags, the relevance of the tag "flower" is $(11 - 2 + 1)/11 \approx 0,909$. The path from image 95 to image 81 is the one shown in Fig. 8. The similarity score between the two images is $1 \times 0,909 = 0,909$ and the recommendation score for image 10 will be $0,909 \times 5 = 4.545$.

Table 8. Use case B: results for user similarity-based recommendations

Score	Image list
4	10

For user similarity, the system obtains only one recommendation (see Table 8). Let N_1, N_2, N_3, N_4 and N_5 be the 5-nearest neighbors of B. The Pearson correlation coefficient values for these neighbors are 0,866, 0,822, 0,799, 0,76, and 0,739, respectively. Image 10 received ratings of 4 from N_1, N_2, and N_3, a rating of 3 from N_4, and a rating of 5 from N_5, thus its score is $(5 \times 0,739 + 4 \times (0,866 + 0,822 + 0,799) + 3 \times 0,76)/(0,866 + 0,822 + 0,799 + 0,76 + 0,739) \approx 4$. It is worth mentioning that the next image in the list had a score of 3,60, fairly distanced from this one.

Since in this case there are no recommendations from the last category, in the ideal case the system would have taken 5 from each of the other two. But, because there is only 1 recommendation from the second category, it will compensate by taking more from the first, the final selection being:

Table 9. Feedback of user B on the recommendations given to them

Image	81	16	64	33	12	79	1	82	61	10
Feedback	5	4	3	3	4	2	2	2	2	2

Fig. 9. Recommendations for user B

- 9 images from the image similarity list: 81, 16, 64, 33, 12, 79, 1, 82, 61;
- 1 image from the user similarity list: 10.

This list of results (shown in Fig. 9) was given to user B for feedback rating. Their response is listed in Table 9. Note that the images with the four lowest ratings for image similarity also received the lowest ratings in the user feedback. This set of recommendations was not as good as the previous one because the user likes few images of the ones they rated and the most similar neighbors had few images with a high rating in common. Nevertheless, half of the recommendations were good ones.

5 System Validation

The system was verified automatically, using the set of 78 users that have rated all 100 images. We have also tested the system on a set of 15 test users, different from the ones in the training set. As criteria for evaluation, we used *accuracy* - what percentage of the recommendations are images the user likes (i.e. the user would give them a rating of 4 or 5). If R is the set of recommended images, the accuracy is given by (8).

$$accuracy(u, R) = \frac{|\{i \in R \mid rating(u, i) \geq 4\}|}{|R|} \tag{8}$$

The accuracy results for both evaluation methods are shown in Fig. 10.

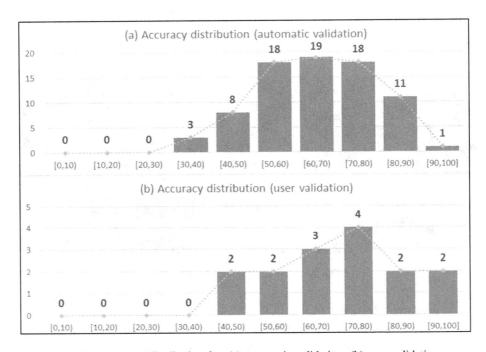

Fig. 10. Accuracy distribution for: (a) automatic validation; (b) user validation.

Automatic validation: Using the training set of 78 users, we have performed a cross validation of our system by taking each user and, in turn, making a random selection of their ratings, then comparing the recommendations of the system against the ratings that we previously discarded. The size of the selection was also random, between 8 and 30. We did 10 runs and, for each user, retained the average over all runs for each of the three evaluation criteria.

We believe the automated validation showed that our system works well. For almost 86 % of cases at least half of the recommendations were good, the bad recommendations showed small faults, most of them between 0,1 and 0,25, and none of them greater than 0,35, while the real rating mean was in 77 out of 78 (98,72 %) cases greater than 3 and in 60 out of 78 (76,92 %) cases greater than 3,5, the average being 3,709.

User-based validation: The output of the system was also tested by 15 users different from the ones in our training set. They were each given a list of 10 recommendations and offered feedback in the form of ratings on the same scale and with the same meaning as the one used when gathering the training data. Although there is little data compared to the automatic validation, the results are alike and prove that the system behaves in a similar manner.

For 86,67 % of cases, the accuracy is at least 50 %. The values for fault range up to 0,5, but most of them are between 0,1 and 0,3. The real rating mean was in 14 out of 15 (93,33 %) cases greater than 3 and in 12 out of 15 (80 %) cases greater than 3,5, the average being 3,78.

The user validation of our system proves that its behavior is consistent with the one shown by the automatic validation. There are some particularities in the values obtained that are most probably influenced by the small amount of data and by the fact that we only repeated the process once. However, the results are satisfactory and prove that the recommendations provided by the system are mostly good ones.

6 Conclusions

We have created an image recommendation system which uses similarity scores for items and users, combined with social profiling for diversity. It provides at least 50 % good recommendations in about 86 % of cases, very few lists of recommendations have a rating mean of less than 3, and about 76–80 % have a rating mean of over 3,5. We evaluated our system automatically and by asking for user feedback, both methods indicating a consistent behavior.

The system was built using a small quantity of data, focusing on experimentation and the choices to be made between several possible approaches. It would be very interesting to examine the possibility of adapting it to work on a larger scale in [10]. One possible approach is to use a distributed system, running in the cloud [11], and using Map Reduce to compute results [12]. We are currently working on gathering more information, by using a collection of at least 1000 images and asking for ratings from a group of about 500 students.

Acknowledgement. The research presented in this paper was funded by the project MUCKE (Multimedia and User Credibility Knowledge Extraction) [10], number 2 CHIST-ERA/01.10.2012.

References

1. Rajaraman, A., Leskovec, J., Ullman, J.D.: Mining of Massive Datasets. Cambridge University Press, New York (2011)
2. Herlocker, J., Konstan, J., Borchers, A., Reidl, J.: An algorithmic framework for performing collaborative filtering. In: Proceedings of the 22nd Annual International ACM SIGIR Conference on Research and Development in Information Retrieval, August 1999
3. Melville, P., Mooney, R.J., Nagarajan, R.: Content-boosted collaborative filtering for improved recommendations. In: Proceedings of the Eighteenth National Conference on Artificial Intelligence (AAAI 2002), Edmonton, Canada, July 2002, pp. 187–192 (2002)

4. Laic, A., Iftene, A.: Automatic image annotation. In: Proceedings of the 10th International Conference "Linguistic Resources and Tools for Processing the Romanian Language", Craiova, 18–19 September 2014, ISSN 1843-911X, pp. 143–152 (2014)
5. Stanford CoreNLP. http://nlp.stanford.edu/software/corenlp.shtml
6. WordNet 3.0 Reference Manual. http://wordnet.princeton.edu/wordnet/documentation
7. A glossary of WordNet terms. http://wordnet.princeton.edu/wordnet/man/wngloss.7WN.html
8. The Neo4j Manual v2.1.2. http://docs.neo4j.org/chunked/milestone
9. Boslaugh, S.: Statistics in a Nutshell. O'Reilly Media, Inc., Sebastopol (2012)
10. Bierig, R., Serban, C., Siriteanu, A., Lupu, M., Hanbury, A.: A system framework for concept- and credibility-based multimedia retrieval. ICMR **2014**, 543 (2014)
11. Pallis, G.: Cloud computing - the new frontier of internet computing. IEEE Internet Comput. **14**(5), 70–73 (2010)
12. Dean, J., Ghemawat, S.: MapReduce: simplified data processing on large clusters. In: OSDI 2004: Sixth Symposium on Operating System Design and Implementation, San Francisco, CA, December 2004

Extracting and Linking Content

Discovering Semantic Relations Within Nominals

Mihaela Colhon[1(✉)], Dan Cristea[2,3], and Daniela Gîfu[2]

[1] Department of Computer Science, University of Craiova, Craiova, Romania
mcolhon@inf.ucv.ro
[2] Faculty of Computer Science, "Alexandru Ioan Cuza" University of Iaşi,
Iaşi, Romania
{dcristea,daniela.gifu}@info.uaic.ro
[3] Institute for Computer Science, Romanian Academy – The Iaşi Branch,
Iaşi, Romania

Abstract. We are interested to develop a technology able to discover entities and relations connecting them, as expressed in fiction texts. Deciphering these links is a major step in understanding the content of books. In this study we consider the case of imbricated entities, therefore entities realized at the surface text level by imbricated spans. For this research we use the *QuoVadis* corpus, whose conventions of annotations we describe briefly, same as some statistics on the types of relations, features regarding the relations' arguments and words or expressions functioning as triggers. The approach to recognize the semantic relations is based on patterns extracted from the corpus. The evaluation shows very promising results.

Keywords: Semantic relations · Nominal expressions · Annotated corpus · Anaphora · Annotation conventions

1 Introduction

Extracting relations among entities is an active research area in the field of linked textual data, with applications in the areas of semantic and social web [6], information extraction [26] and text mining. In the last years, the search engines usage for recognizing the relation between two entities has also gained attention [13]. Text mining implies meaningful representation of texts, including encoding of entities and relations occurring between them.

The field touches the very essence of the deep understanding of natural language. As language exhibits a huge diversity of expression for entities as well as relations connecting them, any consideration here should be made on a firm representational ground. Thus, the main major preoccupation for NLP technologies related to text mining (the enlarged field of the old information extraction area) goes in the following directions: 1. finding sound representations at a conceptual level; 2. decoding language onto this representation, and 3. mimicking the reasoning capacity of humans, which is manifested in our skills to understand and make use of the language in real life.

© Springer International Publishing Switzerland 2016
D. Trandabăţ and D. Gîfu (Eds.): EUROLAN 2015, CCIS 588, pp. 85–100, 2016.
DOI: 10.1007/978-3-319-32942-0_6

In this study we are mainly concerned with parts of the second topic, namely to find ways to map the huge diversity of natural language expressions onto sound conceptual level representations. Our annotations are meant to show simultaneously the constructions at the basic language level and the equivalent encodings at the knowledge representation level (entities and relations between them). If, in general, entities are of a very diverse nature: persons, animals, places, organizations, crafts, objects, ideas, events, moments of time, etc., in this study we are concerned only with persons, gods and any groupings of them. Among the extremely large class of semantic relations that a text could express, we decipher four types: anaphoric (or referential) relations (when the interpretation of one entity mention is dependent on the interpretation of a previous one), affectional relations (when a certain feeling or emotion, is expressed in the interaction of characters), kinship relations (when family relationships are mentioned, sometimes composing very complex genealogical trees), social relations (when job hierarchies or social mutual ranks are explicitly remarked) [5].

Syntactically, the text realization of entities is nominal phrases (NPs). However, NPs have sometimes recursive structures, such that one NP may include one or more other NPs. Since to each NP at the textual level, an entity is mapped at the representational level, the recursiveness of the NP chunks is reflected by relations between corresponding entities. Some examples are: University of Washington (mentioning an institution, a location and the positioning of the institution in that location) or mother of the child (including two entities of type person: the mother of the child and the child itself and a kinship relation linking them). To reflect the surface structure, these entities are often referred to as nested or imbricated [8]. It should be noted that imbricated NPs have always separate heads and there are not NPs that intersect and are non-imbricated [5].

In this paper we propose a method for automatic recognition of semantic relations that are mentioned within imbricated entities. The method uses lexico-syntactic patterns extracted from a training corpus, which are constructed by exploring the complex annotations of the lexical resource.

The paper is organized as follows. In the following sections we give an overview of the similar studies in name entity recognition, relations discovery and nested semantic relations. Section 4 describes our approach for automatic relation recognition within nominals. Section 5 presents evaluation considerations and the last section makes concluding remarks and presents ways of improving this recognition mechanism.

2 Related Work

Most of the designers of corpora dedicated to named entity recognition (NER) usually ignore nested entities by choosing to focus on the outermost entities and proposing in this manner flat entity representations. The widely used MUC-6, and MUC-7 NER corpora, composed of American and British newswire, are flatly annotated. The GENIA corpus [3] contains biomedical named entities, but the JNLPBA 2004 shared task [4], which utilized the corpus, removed all embedded entities for the evaluation [8]. To our knowledge, the only shared task which has included nested entities for evaluation is the SemEval 2007 Task 9 [18], which used a subset of the AnCora corpus.

NLP researchers use corpora annotated with semantic links for training recognition algorithms. Since the 80s, the normalized for nested relations and object-oriented database become objectives for many researchers [21]. A considerable number of studies are concentrated on nested relational normal forms, like: NNF [21], NF2 [22] and NF-NR [16].

Özsoyoglu and Yuan, in NNF (Nested Normal Form), consider that the nested relations are structured as trees, called scheme trees, and introduce a normal form for these relations, called the nested normal form. The representation of nested sets as trees or hierarchies we find it in Nested set model, sometimes with different names like Recursive Hierarchies [12].

NF2 model is, actually, an extension of the classical relational model, which focuses on relation-valued attributes, improving with a reformulation of query operations of the frame model in terms of NF2 algebra operations [22].

NF–NR model removes inconvenient anomalies from a nested relational database schema, such as global redundancies between nested relations [16], considering two approaches: the restructuring the nested relations by applying a set of rules that transform relations NF-NR nested relations and the entity-relationship to NF-NR database design, based on the normal form for ER model [15]. Moreover, the nested relations discusses as a database model, are also research topics for many researchers [11, 17, 23, 24], emphasizing the ability to represent and manipulate complex structures [1].

Furthermore, in literature, a few researchers have focused on introducing imprecise and uncertain information into NF relational database, in so-called the fuzzy nested relational models. We remind a NF database model with null values [14] or the modeling in NF data model of the uncertain null values, set values, range values, and uncertain values, being extended NF algebra on similarity-based [3].

In addition, to define and recognize nested relations, he worked with Relix (RElational database programming Language in UNIX), a system focused on two kinds of data models: domains and relations [10] and the nested relations were built on top of relations and nested queries by allowing the domain algebra to subsume the relational algebra. In this paper, we have focused on syntactic pattern, an approach that has been used. For instance, QBQL syntax, which has built-in set join operations, nested relations provide an alternative.

3 Our Previous Work

Our work continues the research described in [5], and summarized in this section. The essence of the research aimed at building a corpus of annotated entities and semantic relations. The text used was the Romanian version of the novel "Quo Vadis", authored by the Nobel laureate Henryk Sienkiewicz[1]. We have presented in the mentioned paper the marking conventions, designed to incorporate annotations for persons and god type

---

[1] Version translated by Remus Luca and Elena Linţă and published at Tenzi Publishing House in 1991. The aligned passages are searched in the English translation made by Jeremiah Curtin and published by Little Brown and Company in 1897.

entities, including groups, and for relations linking them. The annotation itself was a time consuming and painful process, that run over more than two years.

3.1 Annotating Semantic Relations

Out of the vast variety of relations that could come together with the mentions of characters in fiction texts, we have concentrated on 12 types of anaphoric relations (examples are: `coref`, `member-of`, `part-of`, `has-as-part`, etc.) and almost 30 types of non-anaphoric relations (examples are: `parent-of`, `child-of`, `love`, `friendship`, `hate`, `superior-of`, `inferior-of`, `colleague-of`, etc.). We believe this set of relations covers to a large extend the relational inventory in a fiction text. Each segment of text received by annotators included already basic markings on token/part-of-speech/lemma layers, performed automatically during a pre-processing phase [25]. Based on these, the manual annotation captured the following aspects: notation of mention of entity, the relation's boundaries, the type of relation, its two arguments (all relations are binary) and, where present, the trigger (a word or an expression signaling the relation). All the files contributed by individual annotators were then merged automatically in a contiguous file, IDs of XML elements and their references being re-generated. To this initial level of annotation we added, recently, syntactic dependency data: each token of all sentences has been complemented with its head-word and the dependency relation towards the head.

The corpus thus obtained was used to train a process able to generalize patterns and to identify features for automatic relation discovery. At this stage of the research, we have concentrated only on the automatic recognition of the non-anaphoric relations, also ignoring the identification of entities.

At the text level, the entities are realized[2] as noun phrases whose heads are nouns or pronouns. These constructions may include, besides their heads, also modifiers, such as determiners, adjectives, numerals, genitival constructions, or prepositional phrases. However, we imposed the constraint that noun phrases do not extend over relative clauses.

Considering the relative positioning in the text of the spans of text that realize the entities forming the two arguments of the relations, they could intersect or not. If they intersect, then (empirical evidence show that) they are necessarily nested (imbricated) and the convention for the direction of the relation is to consider as FROM the larger entity and as TO the nested entity [5]. For instance (here and below, entities are marked in square brackets and triggers in angular brackets):

```
1:[<copilul> drag 2:[al celebrului Aulus]] (in the English ver-
sion, 1:[a dear <child> of 2:[the famous Aulus]])
```

---

[2] The term "realize" is the one used in Centering [9].

In this example, the FROM entity noted here with [1] imbricates the TO entity denoted by [2] and there is a child-of relation between [1] and [2]. The trigger here is <copilul> (EN: <child>).

In this study, we explore only the relations occurring between nested entities.

3.2   The Inventory of Semantic Relations

As noted in the literature [7], there is not an universally accepted list of semantic relations to be considered between nominal groups. Different teams of researchers consider their particular lists, very much dependent on the domain of the analyzed texts of the application envisioned. In [5] the semantic relations marked in the *QuoVadis* corpus are grouped in four classes, and the same classification applies to imbricated entities. Some examples follow:

– relations of the class AFFECT:

```
1:[a <prietenului>, amicului și confidentului 2:[lui Ne-
ro]]   =>   AFFECT.   friend-of   (EN:   1:[of   2:[Nero]'s
<friend>, companion, and suggester]).
1:[<favoritul>  2:[împăratului]]   =>   AFFECT.loved-by  (EN:
1:[2:[Cæsar]'s first <favorite>] ).
1:[fosta <amantă> 2:[a lui Nero]] => AFFECT.rec-love (EN:
1:[the former <favorite>2:[of Nero]] ).
1:[unui <credincios> 2:[al "Mielului"]] => AFFECT.worship
(EN: 1:[a <confessor> 2:[of the "Lamb"]]).
```

– relations of the class KINSHIP:

```
1:[propriul  2:[lor]  <copil>]   =>   KINSHIP.child-of  (EN:
1:[their 2:[own] <daughter>]).
1:[scumpii  2:[săi]  <nepoți>]   =>   KINSHIP.nephew-of  (EN:
1:[2:[his] dear <nephews>]).
1:[<părinte>  2:[al  zeilor]]   =>   KINSHIP.parent-of  (EN:
1:[<father 2:[of the gods]] ).
1:[niște  <frați>  2:[ai  tăi]]   =>   KINSHIP.sibling  (EN:
1:[thy 2:[own] <brothers>]]).
1:[nefericita  <soție>  2:[a  lui  Zethos]]   =>   KIN-
SHIP.spouse-of (EN: 1:[the unhappy <wife> 2:[of Zethos]]).
1:[furtunos <urmaș> 2:[al consulilor]] => KINSHIP.unknown
(EN: 1:[mad <descendant> 2:[of consuls]]).
```

– relations of the class SOCIAL:

```
1:[<tovarăşul> 2:[lui Petru]] => SOCIAL.colleague-of (EN:
1:[2:[Peter's] <companion>]] ).
1:[<adversarului>   2:[său]   greoi]   =>   SOCIAL.in-
competition-with (EN: 1:[2:[his] heavy <antagonist>]]).
1:[a   tuturor   <sclavilor>   2:[prefectului   Pedanius
Secundus]] => SOCIAL.inferior-of (EN: 1:[all the <slaves>
2:[of the prefect Pedanius Secundus]]).
1:[cei   mai   aprigi   <duşmani>   2:[ai   Romei]]   =>   SO-
CIAL.opposite-to (EN: 1:[2:[of Rome's] most inveterate
<enemies>]]).
1:[<comandanţi> de 2:[cohorte]] => SOCIAL.superior-of (EN:
1:[2:[pretorian] <leaders>).
```

– relations of the class REFERENTIAL:

```
1:[<unul dintre> 2:[sclavi]] => REFERENTIAL.member-of (EN:
1:[one of my 2:[slaves]]).
1:[Apostolul cu 2:[<barba> argintie]] => REFERENTIAL.has-
as-part (EN: 1:[Apostle with his 2:[silvery beard]])3.
1:[<numele>   de   2:[roman]]   REFERENTIAL.name-of   (EN:
1:[2:[Roman] name]).
1:[<faţa> 2:[ei] tristă, dar senină] REFERENTIAL.part-of
(EN: 1:[2:[her] face, pensive, but mild]).
1:[un <grup> dintre 2:[celelalte slugi]] REFEREN-
TIAL.subgroup-of (EN: 1:[a crowd of 2:[other serv-
ants]]).
```

4   Automatic Recognition of Semantic Relations

Syntactically, the text realization of entities is nominal phrases (NPs). As noted in literature, entity identification and relation extraction are two separate tasks, the second one actually following the first one [20]. We relied, in our study, on the manual annotations for entities in the *QuoVadis* corpus, and focused our attention strictly on the automatic identification of the semantic relations between nested entities. In the following section, we describe two approaches, one using morpho-syntactic information, the other - dependency data, i.e., conforming to [19], asymmetrical functional relations between pairs of words, considered head and modifier.

4.1   Morpho-Syntactic Patterns

A collection of morpho-syntactic patterns has been extracted from the occurrences of imbricated relation spans belonging to the training corpus. Sequences of items covering the whole span of the relation sharing similar morpho-syntactic structures can be considered candidates for the corresponding type of relation. No lexical information is considered here yet, although we are aware that it has an important role in identification of relations. The main scope of this approach is to generalize the syntactic patterns found in the corpus under the same relation realization in order to discover similar sequences, instantiated or not in the corpus, which could belong to the same relation type. To allow a larger flexibility, generalizations of patterns can also be easily mastered by including wildcards.

Let us consider the following example:

```
1:[2:[împăratul    însuşi],      şi    3:[preoţii],      şi
4:[vestalele],  şi  5:[senatorii],  şi  6:[cavalerii],  şi
7:[poporul]]  (EN: 1:[2:[Cæsar  himself]  bet; 3:[priests],
4:[vestals],  5:[senators],  6:[knights]  bet; 7:[the popu-
lace] bet).
```

Here, the entity on the position of FROM argument, [1], is in a REFERENTIAL. has-as-member relation with the entity [2] and with a REFERENTIAL. has-as-subgroup relation with each of the entities [3–7]. The morpho-syntactic structure of the text corresponding to the entity [1] shows an enumeration of noun phrases separated in the text by the connectors Cc (for conjunction) and COMMA (for comma):

```
1:[2:[Ncmsry    Dh3ms]   COMMA   Cc   3:[Ncmpry]    COMMA   Cc
4:[Ncfpry]  COMMA  Cc  5:[Ncmpry]  COMMA  Cc  6:[Ncmpry]  COMMA
Cc  7:[Ncmsry]].
```

At this stage of our study, morpho-syntactic patterns are only used to detect REFERENTIAL relations realized by enumeration sequences, from an external PERSON-GROUP entity to imbricated PERSON and PERSON-GROUP entities. However, as can be noticed from this example, without lexical information that identifies the semantic class of inner entities no distinction can be made among the class of relations that share the same general structure of the two arguments, namely FROM being a group entity and TO being different types of components: member, part or sub-part. These relations are, correspondingly: has-as-member, has-as-part and has-as-subgroup.

4.2 Lexical-Dependency Patterns

In this approach, the text of the sentence that includes the span of the relation is extracted and processed with a Dependency Parser and the resulted dependency links relating the arguments' spans are put in evidence. Here, patterns are generated as a sequence of typed dependencies linking the lemma of the head of the inner entity to the lemma of the head of the outermost entities. For instance, in the following example, where heads are underlined, a KINSHIP.unknown relation is established between the entities [1] and [2], triggered by <urmaș> (EN: <descendant>), which is the head word of entity [1].

```
1:[furtunos <urmaș> 2:[al consulilor]]  (EN: 1:[mad <de-
scendant> 2:[of consuls]])
```

At the dependency level there is a substantive attribute (a.subst) relation linking the head of [2], *consulilor* (*consuls*), with the head of the entity [1], *urmaș* (*descendant*), as is illustrated in Fig. 1.

An empirical analysis of the cases encountered in the training corpus resulted in the following steady observations, put here as hypotheses:

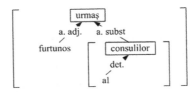

Fig. 1. The dependency parse tree for the text *furtunos urmaș al consulilor* (EN: *mad descendant of consuls*) – example for the a.subst relation

Hypothesis 1: The outer-inner imbrication of the text spans corresponding to the two arguments of an imbricated semantic relation is copied also on the dependency tree, where the tree structure corresponding to the inner span is a sub-tree of the one corresponding to the outer span.

Hypothesis 2: If the direction of a relation in the syntactic tree is considered from a word towards its dependency head (parent), then there is always a path of dependency relations linking the NP head of the inner entity to the NP head of the outer entity.[3]

All entities in our corpus are realized by nominal phrases (head being the central noun of the group) or by pronominal phrases (head being the central pronoun). Three types of dependency relations link the head of the inner entity to the head of the outer entity.

– substantival adjunct (a.subst) – when the head of the inner entity is a noun. The example in Fig. 1 shows the dependency tree for 1:[furtunos <urmaș> 2:

---

[3] Care should be taken when using the term 'head': in the context of the surface text, we say that a group (for instance, an NP) has a head; in the context of a dependency tree, we say that each word of a sentence (excepting the main verb), has a head. When confusion could occur, we will use the expressions 'NP head' for the first case, and 'dependency head' – for the second. As such, an NP head is a word belonging to an NP and, seen as participant in a dependency tree, itself has a dependency head, which is external to the NP.

[al consulilor]], where a KINSHIP.unknown relation holds between the outer and the inner entities.

- adjectival adjunct (a.adj) – when the head of the inner entity is a numeral or an adjectival pronoun (in this case there is an agreement in gender, number and case between the two heads). Figure 2 shows the construction 1:[<surorii> 2: [sale] mai mari], which has the morpho-syntactic pattern 1:[<Ncfsoy> 2: [Ds3fsos] Rg Afpfprn].

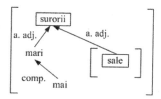

Fig. 2. The dependency parse tree for the text *surorii sale mai mari* (EN: *to her elder sister*) – example for the a.adj relation

- pronominal adjunct (a.pron) – when the head of the inner entity is a pronoun (and there is no agreement between the two heads). Figure 3 shows a KINSHIP. nephew-of relation within the construction 1:[doi<nepoți>de- 2:[ai săi]], which has the morpho-syntactic pattern 1:[Mcmp-1 Ncmprn Sp 2: [Tsmpr Ps3mp-s]].

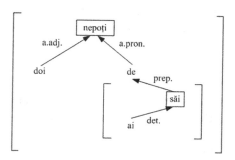

Fig. 3. The dependency parse tree for the text *doi nepoți de ai săi* (approx. EN: *two of his nephews*) – example for the a.adj relation

Hypothesis 3: The trigger of the semantic relation includes (if it is not identical with) the NP head word of the outer entity.

Now, considering the case of two imbricated NPs at the text level, a corollary could be drawn out of the first two hypotheses:

Corollary: A semantic relation between two imbricated entities corresponds to the inverse of a path of dependency relations linking the NP heads of the two entities.

Our recognition algorithm exploits this observation and the lexicalization of the trigger. We used this corollary to prepare an external resource that helped in the recognition process, a process that can be described as here:

- a list of triggers, under the form of lemmas, was created for each type + subtype of a semantic relation identified in the corpus (Appendix A lists the identified triggers in the *QuoVadis* corpus, corresponding to the semantic relations annotated between nested entities);
- a dependency parser was run on the relation spans (the outer entities) on the entire corpus. As a result, the dependency trees of these sub-sentences were generated, the head words of the larger and inner entities were localized, and the dependency path linking the two heads were extracted;
- out of the paths extracted at step 2, those corresponding to annotated outer-inner entities linked by semantic relations were selected and this list was sorted by the semantic relation's type + subtype.

The resources resulted at steps 1 and 3 were merged and grouped by type + subtype of the semantic relations, resulting in a mapping function: $T \rightarrow 2L \times P$, where T is a combination type + subtype, L is a list of trigger lemmas and P is a list of paths of dependency relations. Moreover, the elements of L and P were sorted from the most frequent to the less frequent for each type + subtype of T.

With this resource, the recognition algorithm is an inverse function from $L \times P$ to T, trying to identify that combination type + subtype of a semantic relation which corresponds to the highest product of relative frequencies of the pair (l, p), where l is the lemma of the outer entity and p is a path including or equal to the one linking the inner entity to the outer entity on the dependency sub-tree. This algorithm indeed recognizes the relation type based on the trigger and the dependency structure.

5   Evaluation

The *QuoVadis* corpus contains 22,303 annotated relations, out of which 1,240 occur between nested entities. These pairs of entities were used for building the auxiliary resources and the patterns.

From the total number of relations annotated in the corpus we have kept 90 % for training and 10 % for testing, using a cross-validations policy, which guaranteed no intersection between training and evaluation sentences. It resulted a number of 1,116 relations used in the training phase (24 AFFECT relations, 182 SOCIAL relations, 153 KINSHIP relations and 757 REFERENTIAL relations) and 124 relations for testing (3 AFFECT relations, 20 SOCIAL relations, 16 KINSHIP relations and 85 REFER-ENTIAL relations).

The evaluation task considered here is this: the system is presented with an entity which nests at least one other entity (extracted from the manual annotations of the corpus) and has to decide whether a relation exists between the marked entities and, if yes, to identify its type and subtype (Table 1).

Table 1. Evaluation scores for the recognition of semantic relations between nested entities

Relation type	# test relations	# correct	Precision	Recall	F-measure
AFFECT	27	23	0.9	0.86	0.87
KINSHIP	169	147	0.94	0.87	0.9
SOCIAL	202	155	0.92	0.76	0.83
REFERENTIAL	842	757	0.96	0.9	0.93
Total	1240	1082	0.93	0.85	0.88

As it can be seen, maximum precision and recall is obtained for those types which are sparsely represented in the corpus. We suspect that more training data is needed to get relevant figures. Moreover, the lowest F-measure stays with the SOCIAL class, which has the peculiarity of a very rich set of triggers for each subtype. This high diversification makes also less probable the occurrence of the same relation trigger in both the training and the test parts of the corpus.

The most common problems in recognition are caused by the fact that the head of the FROM entity is not found in any set of triggers extracted from the training corpus. Examples of misses are:

- REFERENTIAL.part-of relation in 1:[căpşorul 2:[fetei]] (EN: 1:[2:[the maiden's] head])

- REFERENTIAL.part-of relation in 1:[carne de 2:[copii]] (EN: 1:[flesh of 2:[children]])

- SOCIAL.inferior-of relation in 1:[unui arendaş 2:[al său]] (EN: 1:[a confidant 2:[of Vinicius]])

- SOCIAL.superior-of relation in 1:[păzitor din oficiu 2:[al ostaticei]] (EN: 1:[official guardian 2:[of the hostage]])

- SOCIAL.superior-of relation in 1:[acel pontifex maximus 2:[al creştinilor]] (EN: 1:[that pontifex maximus 2:[of the Christians]])

There are also sequences where the head word of the FROM entity does not help in recognizing the relation, like in this case for REFERENTIAL.has-name relation, where two other relations are considered to occur: REFERENTIAL.subgroup-of and SOCIAL.inferior-of:

1:[doi sclavi din neamul 2:[quazilor]] (EN: 1:[two powerful 2:[Quadi]])

6   Discussions and Conclusions

This paper presents work in progress in the area of recognition of semantic relations that occur within nested textual entities, linguistically expressed as noun phrases. Although our immediate goal was to develop a tool that would help a user to interact with the *MappingBooks* application, the aim is more generous and the results overpass this limited horizon. We have exposed here a corpus developed in a previous project, conventions of annotation at the entities and semantic relations layers, heuristics for the extraction of semantic relations that make use of lexical and syntactic data, and previous results showing already a promising accuracy.

We are aware already of a number of problems that our approach could hide. One of them is, for example, how could enumerations (of elements in a set, or of parts of a whole, or of subsets belonging to a larger set) be distinguished from apposition or enumeration of qualities describing the same entity. An example is the following:

```
    1:[a  <prietenului>... şi confidentului 2:[lui  Nero]]
    (EN: 1:[a <friend> and confident 2:[of Nero]]
```

where both terms of the conjunction refer the same person and an AFFECT. friend-of relation should be recognized between [1] and [2].

Another potential problem is given by the need to go beyond the lexical lists suggested by triggers as cues to distinguish different types of semantic relations, in the need to enhance the recall. Indeed, we are aware that going out of our annotated corpus, to other documents or genres, the recall will deteriorate drastically. To make the tool more reliable, sources that contain semantically related words, as WordNet, are needed. For instance, if the training corpus contains *soţie* (EN: *wife*), it would be good to have also: *soaţă, nevastă, muiere, consoartă, femeie, tovarăşă de viaţă, jumătate, pereche* etc. (synonyms and metaphors of *wife*).

Finally, ambiguities, therefore deteriorating the precision, are induced by triggers having more senses, as here:

```
    1:[<capul> 2:[omului]] (EN: 1:[2:[man's] <head>]), in-
    ducing a REFERENTIAL.part-of relation
```
and here:
```
    1:[<capul> 2:[familiei]] (EN: 1:[<the head> of 2:[the fam-
    ily]])
```

inducing a SOCIAL.superior-of relation. To face this problem, a word sense disambiguation phase should precede detection of relations.

Acknowledgements. This work was partly supported by *MappingBooks*, a project financed by the Romanian Ministry of Education and Research (UEFISCDI) under the Partnerships Programme (PN II Parteneriate, competition PCCA 2013), project code: PN-II-PT-PCCA-2013-4-1878.

Thanks are addressed to Radu Simionescu for performing the dependency parsing of the *QuoVadis* corpus and all members of the NLP-Group@UAIC-FII which contributed to the creation of this corpus.

Appendix A

See Table 2.

Table 2. List of triggers grouped by their semantic relations

Semantic relations	Lemmas of triggers (#occurrences)	Semantic relations	Triggers
AFFECT.fear-of	*teme* (9), *frică* (3), *fior de groază* (1), *teamă* (1), *tremura* (1)	SOCIAL.colleague-of	*tovarăş* (8), *frate* (4), *aliat* (1), *compatrioată* (1), *coreligionar* (1), *oaspăt* (1), *protector* (1)
AFFECT.friend-of	*prieten/ prietenă/ prietenie* (28), *amic* (2), *pare rău* (1),	SOCIAL.in-competition-with	*adversar* (3), *întrece* (2), *bănuială* (1), *concura* (1), *favorit* (1), *rivală* (1)
AFFECT.hate	*ura/ ură/ urî* (10), *a nu iubi* (4), *dispreţui* (1), *duşmănos* (1), *pârî* (1), *răzbuna* (1)	SOCIAL.in-cooperation-with	*ajutor* (1), *tovarăş* (1)
AFFECT.hated-by	*ură, urî* (3)		
AFFECT.love	*iubi* (44), *dragoste* (8), *adora* (7), *drag* (5), *iubit/iubită* (5), *îndrăgosti/ îndrăgostit* (5), *dor* (2), *favorit* (2), *afecţiune* (1), *devotat* (1), *îndurător* (1), *plăcea* (1), *scump*	SOCIAL.inferior-of	*sclav/ sclavă* (33), *poruncă/ porunci* (19), *libert/ libertă* (14), *om* (14), *serv* (9), *adept* (7), *slugă* (6), *apostol* (5), *centurion* (4),

AFFECT.loved-by	(1), *tânji* (1) *iubi* (36), *dragoste* (6), *drag* (5), *favorit* (2), *îndrăgi* (2), *admira* (1), *îndrăgostit* (1), *vrea iubire* (1)		*slujitor* (4), *condus* (3), *credincios* (3), *curtean* (3), *ostatecă* (3), *supune/ supunere/ supus* (3), *ales* (2), *ostaş* (2), *preot* (2), *suit* (2), *arendaş* (1), *asculta* (1), *cohortă* (1), *consul* (1), *discipol* (1), *gardă* (1), *îngenunchea* (1), *învăţăcel* (1), *învins* (1), *legiune* (1), *locţiitor* (1), *ordin* (1), *prosterna* (1), *rob* (1), *servitor* (1), *sluji* (1), *soldat* (1), *sub comandă* (1), *ucenic* (1), *un demn prozelit al lui Christos* (1), *servitor* (1), *victima* (1)
AFFECT.rec-love	*dragoste* (2), *amantă* (1), *îndrăgostită* (1), *iubi* (1)		
AFFECT.upset	*întrista* (1), *părea rău* (1)		
AFFECT.worship	*închina* (2), *ruga* (2), *cinsti* (1), *credincios* (1), *prosterna* (1), *slăvi* (1)		
AFFECT.worshiped-by	*adorată* (1), *adoratoare* (1), *slăvi* (1)		
KINSHIP.aunt/uncle-of	*unchi* (1)		*duşman* (14), *împotrivi* (3), *împotriva* (2), *advers* (1), *bogat* (1), *călău* (1), *deosebit* (1), *învins* (1), *stăpân* (1)
KINSHIP.child-of	*fiu/ fiică* (69), *copil/ copilă* (21), *copie* (5), *băiat* (2), *vlăstar* (2), *fată* (1), *nepot* (1), *urmaş* (1)	SOCIAL.opposite-to	
KINSHIP.nephew-of	*nepot* (8)		
KINSHIP.parent-of	*mamă* (19), *tată* (10), *părinte/ părinte adoptiv* (6), *fiu* (1)		*stăpân/ stăpână* (27), *rege* (8), *mai mare/ mai marele/ mai marii* (7), *comandant* (6), *în frunte* (4), *comanda* (3), *prefect/ prefectură* (3), *conducere* (2), *învăţător* (2), *mare* (2), *zeu* (2), *cârmuieşte* (1), *conduce* (1), *dispune* (1), *domina* (1), *domn* (1), *dumnezeu* (1), *împărat* (1), *imperator* (1), *maestru* (1), *păzitor* (1), *sclav* (1), *superior* (1), *supraveghetor* (1)
KINSHIP.sibling	*frate* (10), *soră* (8), *soţie* (1)	SOCIAL.superior-of	
KINSHIP.spouse-of	*soţie/ soţ* (27), *logodnic/ logodnică* (6), *amantă* (4), *nevastă* (4), *iubită* (1)		
KINSHIP.unknown	*rudă* (9), *concubină* (2), *strămoş* (2), *neam* (1), *strănepot* (1), *urmaş* (1)		

References

1. Abiteboul, S., Scholl, M.: From simple to sophisticated languages for complex objects. IEEE Data Eng. **11**(3), 15–22 (1988). Special Issue on Nested Relations
2. Bibiri, A.D.: An Annotated Corpus of Entities and Semantic Relations. Dissertation thesis presented at The Faculty of Computer Science, "Alexandru Ioan Cuza" University of Iaşi (2014)
3. Buckles, B.P., Petry, F.: A fuzzy representation of data for relational databases. Fuzzy Sets Syst. **7**, 213–226 (1982)
4. Collier, N., Kim, J., Tateisi, Y., Ohta, T., Tsuruoka, Y. (eds.): Proceedings of the International Joint Workshop on NLP in Biomedicine and its Applications (2004)
5. Cristea, D., Gîfu, D., Colhon, M., Diac, P., Bibiri, A.-D., Mărănduc, C., Scutelnicu, L.-A.: Quo vadis: a corpus of entities and relations. In: Gala, N., Rapp, R., Bel-Enguix, G. (eds.) Language Production, Cognition, and the Lexicon. Springer, Heidelberg (2015)
6. Ding, L., Finin, T., Joshi, A., Peng, Y., Cost, R.S., Sachs, J., Pang, R., Reddivari, P., Doshi, V.: Swoogle: a semantic web search and metadata engine. In: Proceedings of the 13th ACM Conference on Information and Knowledge Management (2004)
7. Downing, P.: On the creation and use of English noun compounds. Language **53**(4), 810–842 (1997). doi:10.2307/412913
8. Finkel, J.R., Manning, C.D.: Nested named entity recognition. In: Proceedings of the 2009 Conference on Empirical Methods in Natural Language Processing (EMNLP 2009), pp. 141–150 (2009)
9. Grosz, B., Joshi, A., Weinstein, S.: Centering: a framework for modelling the local coherence of discourse. Comput. Linguist. **2**(21), 203–225 (1995)
10. Hongbo, H.E: Implementation of Nested Relations in a Database Programming Language, Master Thesis, School of Computer Science McGill University, Montreal (1997)
11. Jaeschke, G., Schek, H.J.: Remarks on the algebra of non first normal form relations. In: ACM Symposium on Principles of Database Systems, Los Angeles (1982)
12. Kamfonas, M.: Recursive hierarchies: the relational taboo! The Relational J., October/ November 1992
13. Keller, F., Lapata, M.: Using the web to obtain frequencies for unseen bigrams. Comput. Linguist. **29**(3), 459–484 (2003)
14. Kim, J.-D., Ohta, T., Tateisi, Y., Tsujii, J.: GENIA corpus semantically annotated corpus for bio-textmining. Bioinformatics **19**, 180–182 (2003). doi:10.1093/bioinformatics/btg102
15. Ling, T. W.: A normal form for entity-relationship diagrams. In: Proceedings of 4th International Conference on Entity-Relationship Approach (1985)
16. Ling, T.W., Yan, L.L.: NF-NR: a practical normal form for nested relations. J. Syst. Integr. **4**, 309–340 (1994)
17. Makinouchi, A.: A consideration on normal form of not-necessarily normalized relation in the relational data model. In: Proceedings of 3rd International Conference on VLDB, Japan (1977)
18. Márquez, L., Villarejo, L., Marti, M.A., Taulè, M.: Semeval-2007 task 09: multilevel semantic annotation of Catalan and Spanish. In: Proceedings of the 4th International Workshop on Semantic Evaluations (SemEval-2007) (2007)
19. Mel'čuk, I.: Dependency Theory: Syntax and Practice. SUNY Press, Albany (1987)
20. Năstase, V., Nakov, P., Séaghdha, D.Ó., Szpakowicz, S.: Semantic Relations Between Nominals. Morgan & Claypool Publishers, San Rafael (2013)
21. Özsoyoglu, Z.M., Yaun, L.-Y.: A new normal form for nested relations. ACM Trans. Database Syst. J. **21**(1), 111–136 (1987)

22. Reimer, U.: A frame-based knowledge representation model and its mapping to nested relations. Data Knowl. Eng. J. **12**(4), 321–352 (1989)
23. Roth, M.A., Korth, H.F.: The design of ¬1NF relational databases into nested normal form. In: Proceedings of the ACM SIGMOD International Conference on Management of Data, pp. 143–159 (1987)
24. Roth, M.A., Korth, H.F., Siberschatz, A.: Extended algebra and calculus for nested relational databases. ACM Trans. Database Syst. J. **13**(4), 389–417 (1988)
25. Simionescu, R.: Graphical grammar studio as a constraint grammar solution for part of speech tagging. In: Moruz, M.A., Cristea, D., Tufiş, D., Iftene, A., Teodorescu, H.N. (eds.) Proceedings of the 8th International Conference Linguistic Resources and Tools for Processing of the Romanian Language, pp. 109–118 (2012)
26. Sheth, A.P., Ramakrishnan, C.: Relationship web: blazing semantic trails between web resources. IEEE Internet Comput. **11**(4), 77–81 (2007)

Quality Improvement Based on Big Data Analysis

Radu Adrian Ciora[✉], Carmen Mihaela Simion, and Marius Cioca

Faculty of Engineering, "Lucian Blaga" University of Sibiu, Sibiu, Romania
{radu.ciora,carmen.simion,marius.cioca}@ulbsibiu.ro

Abstract. Big data analysis has become an important trend in computer science. Quality improvement is a constant in current industry trends. In this paper, we present an idea of quality improvement based on big data analysis with the aid of linked data and ontologies in order to implement it in the case of automotive parts production. We consider defective automotive products and try to find the best refurbishment solution for them considering their characteristics. Moreover, we propose to develop a recommender system that is able to give recommendations in order to prevent or to alleviate defects and to provide insights for possible causes that led to these defective parts. This study intends to help direct beneficiaries (public consumer, quality engineers, quality control managers), but also specialists and researchers in the NLP, software engineers, etc.

Keywords: Big data · Linked data · Data modelling · Business intelligence · Quality improvement

1 Introduction

In every production line there is at least one quality checkpoint. In some industries, these checkpoints check all final products. In other industries, products are checked on a sample basis. In general, much of the quality checking operations are made by humans, by visual inspection. This is a very eye stressful operation, which is why the operators have to take frequent breaks in order to avoid eye fatigue. Moreover, having a tired operator can lead to even worse results in quality assurance. In order to recycle or reuse defective parts, firstly an operator has to determine if the part is good or bad. Secondly, it has to determine the part's defects and then to consider either trying to refurbish if this is still possible or consider another final product that this defective part can be reused for. Thirdly, in a production line we have several machines that model the part in question. If anyone of these becomes defective, the whole lot of products shaped by the defective machine may be compromised. Thus, early detection of a defective machine is important in a company's economy. What we are dealing with here is a big data analysis problem, which can be hard to be solved by an individual in a short period of time.

Nowadays, computer vision based inspection systems are more and more seen in the quality assurance processes. Computers are able to store and process a large amount of data in a very short period of time. It is the so called big data analysis [14]. Big data is a concept of large datasets that are not able to be captured, curated and processed within a reasonable amount of time, including analysis of traditionally structured, semi-structured and unstructured data. Structured data has the advantage of being easily

D. Trandabăţ and D. Gîfu (Eds.): EUROLAN 2015, CCIS 588, pp. 101–109, 2016.
DOI: 10.1007/978-3-319-32942-0_7

manageable, queried and analysed. Semi-structured data includes data from sensors, from process control monitoring tools, annotations etc. In order to process this data firstly it needs to be converted into a structured form. This can be achieved straight forward with the aid of annotation and categorization. On the other hand, unstructured data is data that comprises mainly from text generated from emails, transcripts, documents, spreadsheets, but also from images and video sequences. If for written information, annotation and categorization are the necessary step. Image and video processing require additional pre-processing steps, for object identification, isolation and validation based on feature extraction. Having this amalgam of data sources and data types, it is obvious that finding a common ground is a must if we want to be able to exploit correctly and fruitfully the power that lies in this data.

In this paper, we present an idea of quality improvement based on big data analysis with the aid of linked data and ontologies in order to implement it in the case of automotive parts production. We consider defective automotive products and try to find the best refurbishment solution for them considering their characteristics. Moreover, we propose to develop a recommender system that is able to give recommendations in order to prevent or to alleviate defects and to provide insights for possible causes that led to these defective parts.

Thus, we provide a possible solution to enhance production quality, by providing a tool that is capable to analyse virtual data in real-time and provide solutions for problems which arise on a production line.

The paper is structured as follows: Section. 2 puts our work in the context of the existing work in the field, Sect. 3 describes the problem addressed by this paper, Sect. 4 details our approach towards the exposed problem. Finally, Sect. 5 presents our results, the concluding remarks and future development directions.

2   Related Work

Today many manufactured parts can be designed on a computer, visualized on screen, made by computer-controlled machines and inspected by the very same computer - all without human intervention in handling the parts themselves. There is much sense in this computer integrated manufacture (CIM) concept, since the original design dataset is stored in the computer and therefore it might as well be used (1) to aid the image analysis process that is needed for inspection and (2) as the actual template by which to judge the quality of the products [4]. In other words, there is no need for a separate set of templates for inspection criteria, when the specifications are already on the computer.

In [9], Lee introduces the concept of Cyber-Physical Systems (CPS), as an integration between computation and physical processes. They comprise of embedded systems, network monitors and physical processes controls, which are interconnected with the aid of feedback loops, which allow for physical processes to interact with computations and vice versa. The paper emphasis the potential of such systems, although challenges exist, especially concerning the physical components of such systems which have to obey certain qualitative safety and reliability requirements, which do not apply to software engineering. Moreover, the characteristics of object-oriented software components have nothing in common with the ones of physical components in terms of quality assurance. The author concludes that is not enough to improve the design of process, increase the level of abstraction or to verify the design of today's levels of abstraction, but rather one needs to rebuild the level of abstraction of computing, networking and physical subassemblies so that they can be abstracted in a unified way.

The idea is enhanced by Aruväli et all, under the name of Cyber Physical Production System (CPPS) concept, capable of monitoring and assisting the manufacturing process [2]. He introduces the application of Digital Object Memory (DOMe), which creates an intelligent environment of sensors called augmented reality. The basic idea of their paper is that every part of the manufacturing processing, from parts, notes, drawings, tools, raw materials are linked to DOMe information that is semantic information about its provenance and context based rules. This information is then put into a context comprised of three components: a user, an object and a location. An intelligent environment needs to exist in order to monitor and provoke interaction between objects.

Metz et al. propose an event processing engine which was developed in order to monitor and control the manufacturing process in almost real time [10]. This system considers events as being rules on which event patterns can be applied in order to analyse the process evolution. The pattern definition is not an online process, but rather an offline one. Thus, the patterns are not able to adapt to variations of the monitored process and so the offline activity has to be transferred manually to the online system. Their contribution is represented by a methodology which tries to overcome the aforementioned down-side. Thus a rule-based machine learning algorithm is presented as a solution used to adapt the system automatically to new events. Thus the system is able to identify and validate rules using rule induction techniques. On detection of certain predefined patterns the event processing engine executes a sequence of steps indicated by the rule induction manager and at the same time processes the current process background information. These steps are: selection of a suitable background information of the problem; selection of an appropriate rule induction algorithm; discovery, evaluation, validation and generalization of a correct rule and conversion from rule into an event pattern. Thus new patterns can be derived in order to solve new problems.

3   Problem Definition

In this paper, we present a concept developed in order to improve production line quality. In current industrial environment, there is a large amount of information generated in a time quanta. This information cannot be processed in real time using traditional data mining techniques and technologies. We use the concept of big data because of its possibility to analyse data, including unstructured data like texts, images or videos which could not be collected, pre-processed, searched with traditional data mining instruments and its ability to retrieve meaningful patterns and possible solutions for spontaneous problems. Therefore, we can easily say that big data is a much broader and complex form of existing data mining tools and analysis mechanisms in terms of data analysis, data types, analysis speed and maybe most important analysis scope. For instance, if we take the case of defective mechanical automotive parts, with existing analysis methods little can be inferred that isn't already know, about what caused the defect, what can be done to improve the defect and for sure we won't be able to tell when and how probable the next faults will reappear. Moreover, in some cases, bad decisions can be taken based on inconsistent or limited number of options, choices or pre-existing knowledge, and for example, if we are taking orchestration into account pipeline machines, adjustment can be very difficult.

4   Proposed Solution

A possible solution for this problem mitigation is the use of big data. The use of big data can aid to successfully determine the cause of the problem. Apart from giving a useful solution to the current problem, it can also predict future problems based on possible patterns discovered in the past. It can be easily integrated in orchestrated environments where defects can lead to serious production pitfalls because of the defects that perpetuate on the production line, resulting in huge costs.

Big data has the advantage of using fresh sources of information, while traditional data mining techniques use pre-stored data, which can be outdated or irrelevant for the context of the problem. Another bonus card received by using big data is the possibility to analyse current incidents in real-α time.

Having in view these characteristics, big data was considered an optimal solution for quality assurance in production pipelines. We see this as a very good opportunity for exploiting the benefits of big data over existing solutions.

As represented in Fig. 1, the system architecture is made up of four important blocks: the data acquisition block, the pre-processing block, the ontology development block and the ontology querying block.

The data acquisition block is intended as the name suggests for acquiring multiple pieces of information, in different formats from various input sources. This data can be either structured data, for example, product specification files, which can be used as input in the ontology development process. Semi-structured data is data that has some structuring and might or might not require entering a pre-processing phase in which it

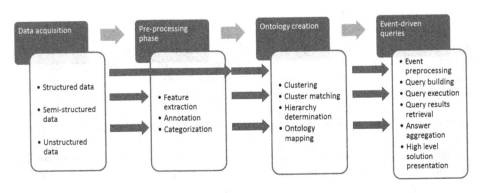

Fig. 1.   System architecture. Information flow

is converted to a more structured format [1]. On the other hand, this pre-processing phase is not the case for unstructured data, as this data is meaningless and useless unless annotation and categorization takes place [6]. Moreover, on moving or still pictures, feature extraction is a must, in order to take full advantage of the rich information which resides in these type of data structures [7].

For the rapid image processing, in order to process the captured images in real time, operations images are compared with predefined templates [3]. If the image fits the template at a certain grade, then the product is considered conformable. First, the determination of the group of pixels, which are of interest and determine the size and structure of the group. Then, we must choose an accurate probability model that can find in a real-α time quanta, a relatively high dimensional pattern [11].

Categorization and annotation are essential operations that need to be added to image processing, in order to develop our ontology. In order to correctly categorize images, a training phase needs to be employed in the pre-processing.

Another important step of this architecture is the ontology creation phase. If this is done properly, the rest will follow naturally, because if the system incorrectly maps a good sample as a bad sample, it will determine the ontology to become inconsistent. Thus, supervised training in the ontology development in the incipient phase is needed, in order to receive expected results. In this phase, information that is pre-processed is clustered into triples, in order to be added to the ontology graph. These clusters are then organised into a hierarchy, going from general information at the root of the hierarchy, towards detailed solutions at the bottom of this hierarchy.

The ontology built can be queried in real time, using SPARQL. The SPARQL queries are triggered based on predefined events and have parameters pre-processed data from sensors, cameras and other sources [13]. The queries return results about the conformity of the products. In case of faults, remedy solutions are suggested.

5   Results

The above solution has been partially implemented in a quality assurance equipment on a production line of an automotive company. Here, the quality of cylinders is assessed. In Fig. 2(a) an image of a cylinder is shown.

Block matching algorithms are usually used in machine vision object detection algorithms. A rising trend has been observed for detection of objects in images where occlusions and clutter exist, so that parts of the objects are missing. Moreover, the object has to be detected even if there are a lot of noise on the object itself. Industrial applications require from object detection algorithms that they are robust to non-linear illumination changes.

Various difference measures which have different mathematical properties and different computational properties had been used to find the best match. The most popular similarity measures based on grey levels are the sum of absolute differences (SAD), the sum of squared difference (SSD), and the normalized cross correlation (NCC). NCC measure is more accurate but, on the other hand it is computationally expensive. It is more robust than SAD and SSD under uniform illumination changes, so the NCC measure has been widely used in object recognition and industrial inspection such as in [5].

But these grey level algorithms proved not to be robust when they need to deal with non-linear illumination changes. In the following we propose a method that can find objects in the presence of occlusion, noise, clutter and non-linear illumination.

An important characteristic of objects in the images that is robust to illumination changes are edges. This is the reason why they are successfully used in robust template matching algorithms. The only difficulty that the use of edges rises is the selection of an appropriate threshold for their segmentation Fig. 2(b). Since threshold selection cannot be perfect, occlusion, noise and clutter handling is essential. Another strategy based on edge segmentation is the derivation of salient points in the image and match them against salient points in the template Fig. 2(c). These salient points can be extracted from the image directly without extracting edges first.

We also investigated another class of algorithms of edge matching which is based on the distance from edges of the template to the edges of the image. The metric we used was the mean squared edge distance (SED). Since it is based on edges, this algorithm is robust to non-linear illumination changes. If clutter is taken into account, extra edges, decrease the distance to the closest edge in the search image, thus the algorithm proved resistant to clutter. But, if edges are missing in the searched image, the distance from

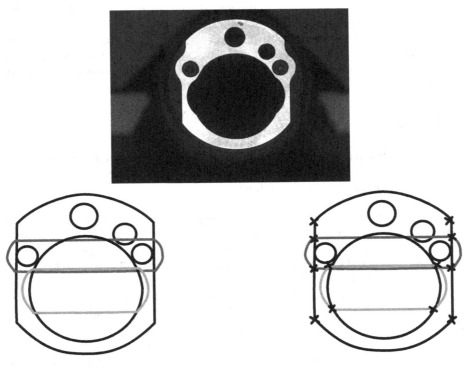

Fig. 2. (a) Image of a cylinder; (b) Edges of the cylinder in (a); (c) Salient points resulting from segmentation in lines and circles.

the temples edges to the closest image edge can become quite big, thus making it difficult to find a correct match.

To overcome these problems, a solution based on Hausdorff distance of two point sets was implemented. If we call the edge point in the template T and the edge points in the image I, then the Hausdorff distance of the two point sets is given by

$$H(T, I) = \max(h(T, I), h(I, T)$$

where

$$h(T, I) = \max_{t \in T} \min_{i \in I} \|t - i\| \tag{1}$$

and $h(I,T)$ is symmetrically defined. Thus, the Hausdorff distance consists of determining the maximum of two distances: the maximum of the distance between the edges of the template and the closest edges in the image and the maximum of the distance between the edges of the image and the closest edges of the template. It can be noted easily that in order to obtain a small distance, every point on the edges of the template has to be very close to a point belonging to the edges of the image and vice versa. Thus, the Hausdorff distance is sensitive to clutter and noise. Responsible for these poor results is the fact that in Eq. (1), the maximum of the distance between the edges of the template and the edges of the image is calculated. In order to have better results in case of occlusions, instead of calculating the longest distance, there can be calculated a distance of a different rank, for example, the f-th longest distance, where $f = 0$ meaning the longest distance. Thus, the Hausdorff distance can correctly treat $100f/n$ % of the occlusions, where n represents the number of points which belong to the template. In order for the Hausdorff distance to correctly treat noise, the $h(T,I)$ function needs to be modified so that it uses the r-th longest distance. Hausdorff distance can be calculated using distance transforms: one for the region of edges in the image and one for every pose of the edges of the template. Thus a large number of distance transforms have to be calculated offline, which is very expensive in terms of memory, or the model's distance transforms can be calculated during the search, but this was neither an option as is computational intensive. In [12] Rucklidge proposes several options for reducing the computational cost, including the prunning of search space of regions that cannot contain the template. Moreover, a hierarhical splitting of the search space is proposed. This is somehow similar with the usage of image piramids except for the fact that the solution provided in [12] just splits the the search space and it does not scale the image and the template, thus remaining relatively slow. Still we managed to take advantage of image piramids, by insipring from [8].

The only drawback of Hausdorf distance is that at moderate levels of occlusion, many false positives instances of the template will be detected in the image. But in our case this is irrelevant as the field of view cannot be obturated, as it is in a controlled environment and so we can ignore this negative aspect of the algorithm. All in all, we consider de use of Hausdorf distance satisfactory for our needs.

After the part in question is successfully detected and compared with the template, our application checks its compliance, by comparing the Hausdorf distances of the golden sample to the sample in the image. There are two validation steps involved. The first validation step checks that sum of all Hausdorff distance does not exceed a certain threshold which is specified by the customer. The second validation step checks that the Hausdorff distance of each salient point in the template to its coresponding point in the image is lower than a predifined threshold. If either of the validation conditions isn't fullfilled, the product is declared defective.

Human intervention can increase the system's ability of finding compliant products, by indicating the machine that a certain defective declared part is compliant. Thus, the machine learns that any part that has characteristics better then the one validated by human operator, will be considered compliant.

6   Conclusions and Future Work

We have described a system that should be self-healing and fault tolerant, able to find solutions to repair defective parts, if possible and if not to mark them as totally compromised.

A major purpose of inspection systems is to take instant decisions on the compliance of products. Related to this purpose are the often fluid criteria for making such decisions and the need for training the system in such a way that the decisions that are would be at least as good as those that would have arisen with human inspectors. In addition, training schemes are valuable in making inspection systems more general and adaptive, particularly with regard to change of product.

If we consider the extension of this architecture, orchestration would be next thing that can be achieved with such a system. With the aid of ubiquitous computing and interconnected machines in the production line, such a system should be able to answer event-driven queries, also, in pipeline architecture and so to be able to adjust the "orchestra" if it fails to perform correctly. Moreover, based on the symptoms that this system can detect, it should also be possible to predict future defects, before they happen, based on small variations of machines parameters, which we could call rumours.

These rumours are the result of self-learning from monitoring and control of the production line, based on history, rule induction and events capturing, curation and analysis. Thus, from observing, recording and understanding the manufacturing process – mapping all this knowledge into the ontology, event-driven queries can be run and should produce results that otherwise couldn't have been obtained using traditional data mining techniques. Therefore, an event driven querying module would be able to decide based on it input from sensors, camera, logs etc. if a fault appeared in the system or not. With the aid of big data analysis. This event driven just-in-time querying system would be able to make its own correct decisions regarding production quality, defects detection and fault recovery.

References

1. Abitebou, S.: Querying semi-structured data, database theory - ICDT 1997. In: 6th International Conference, Delphi, Greece, 8–10 January 1997 (1999)
2. Aruväli, T., Mass, W., Otto, T.: Digital object memory based monitoring solutions in manufacturing processes. Procedia Eng. **69**, 449–458 (2014)

3. Chattopadhyay, T.: Applications of image processing in industries. CSI Commun. pp. 8–11 (2012)
4. Davies, E.R.: Computer and Machine Vision: Theory, Algorithms, Practicalities (2012)
5. Du-Ming, T., Chien-Ta, L.: Fast normalized cross correlation for detect detection. Pattern Recogn. Lett. **24**, 2625–2631 (2003)
6. Feldman, R., Sanger, J.: The Text Mining Handbook, Cambridge University Press (2007)
7. Kaplan, A., Mamou, J., Gallo, F., Sznajder, B.: Multimedia feature extraction in the SAPIR project. In: Proceedings of GSCL (2009)
8. Kwon, O.-K., Sim, D.-G., Park, R.-H.: Robust Hausdorff distance matching algorithms using pyramidal structures. Pattern Recogn. **34**(10), 2005–2013 (2001)
9. Lee, E.A.: Cyber physical systems: design challenges. In: 11th IEEE International Symposium on Object Oriented Real-Time Distributed Computing (ISORC), pp. 363-369 (2008)
10. Metz, D., Karadgi, U.M., Grauer, M.: Self-learning monitoring and control of manufacturing processes based on rule induction and event processing. In: eKNOW 2012: The Fourth International Conference on Information, Process, and Knowledge Management, pp. 88–92 (2012)
11. Pal, C.J.: A Probabilistic Approach to Image Feature Extraction, Segmentation and Interpretation, PhD thesis (1999)
12. Rucklidge, W.J.: Efficiently locating objects using the Hausdorff distance. Int. J. Comput. Vis. **24**(3), 251–270 (1997)
13. Suciu, D. Query decomposition and view maintenance for query languages for unstructured data. In: Proceedings of the 22nd VLDB Conference, Mumbai, pp. 227–238 (1996)
14. Won, H.R.: Rumors on big data and how to utilize big data. In: Finance (2014). http://www.lgcnsblog.com/it-trend/rumors-on-big-data-and-how-to-utilize-big-data-in-finance1/ (Accessed 26 May 2015)

Romanian Dictionaries. Projects of Digitization and Linked Data

Mădălin Ionel Patraşcu[1,2], Gabriela Haja[1], Marius Radu Clim[1], and Elena Tamba[1(✉)]

[1] "A. Philippide" Institute of Romanian Philology,
Romanian Academy, Iaşi, Romania
madalin.patrascu@gmail.com, gabi.haja@gmail.com, marius.clim@gmail.com,
isabelle.tamba@gmail.com
[2] Faculty of Computer Science, Alexandru Ioan Cuza University, Iaşi, Romania

Abstract. In the context of globalization and of interest for linked data, Romanian lexicography tries to harmonize to this trends by aligning its resources and adapting to the necessities of a diversity of users. The lexicographic tradition of the Romanian language passed through various periods, from glosses and small bilingual dictionaries, written in Slavonic alphabet (17th–19th century), to scholar dictionaries from the 20th century, written in Latin alphabet. This tradition was highlighted by different projects, some of them presented in this article, and these projects will continue to emphasize the Romanian language features in order to make accessible the Romanian language for the users and to offer the public research materials and resources of the Romanian culture.

Keywords: e-lexicography · Digitization · Romanian dictionaries · Alignment · Linked data · Machine readable dictionary

1 Introduction

The lexicographic tradition regarding the Romanian language is related to the historic context of the cultural evolution from the Central and South-Eastern Europe, concomitant with the emergence of linguistic identity, as an effective form of shaping national identities. Thus, the first dictionary of the Romanian language is *Dictionarium valachico-latinum* (see [1]), drawn up in the middle of the 17th century and remained as manuscript until 2008 (see [1]), opening the series of bilingual and polyglot dictionaries that made the glory of the following centuries. The monolingual dictionaries appear later and the most important of them is *Dicţionarul limbii române (DA & DLR, The Romanian Language Dictionary)* (see [2,3]), elaborated under the aegis of the Romanian Academy during the period 1905–2010. The first concrete results of the digitizing process of dictionaries (by various ways of transposing the printed form into electronic format) appeared only in the last two decades, and the process is still in progress.

This article presents the projects developed specifically for creating electronic resources required for the elaboration of the second edition of DLR, and those which highlight the dictionaries from the DLR Bibliography (see [4]).

D. Trandabǎţ and D. Gîfu (Eds.): EUROLAN 2015, CCIS 588, pp. 110–123, 2016.
DOI: 10.1007/978-3-319-32942-0_8

2   Encoding Standards for Dictionaries

The lexicographic creations can be stored in the digital environment under various forms. SGML and XML are the most common types of encoding. In addition to these methods of computerized storage, there are also digital format versions related or derived from those mentioned above. Usually, the formats HTML, PDF, TXT, DOC, EPUB etc. are used in the virtual space in order to facilitate the access to lexicographic resources. Nevertheless, it is preferable to use the first encoding version for Machine readable dictionaries and the information can be extracted and transposed into the formats mentioned above, according to necessities.

SGML – *Standard Generalized Markup Language* – was the initial coding solution of Machine readable dictionaries and was used during the '80s and '90s. The passing to XML was forced by the liberty in structuring information and creating annotations, which this format was permitting. Though, this liberty has generated a diversity of encoding models to Machine Readable dictionaries, with the help of XML.

In order to standardize the structure of an electronic lexicographic work and to generate similar structure for other types of dictionaries, the TEI consortium – *Text Encoding Initiative* – has created a standard for this purpose. The new regulation represents the necessary frame for elaborating inter-operable Machine Readable dictionaries.

TEI:P5 Guidelines[1] has reached the 2.8.0 version and analyses in chapter *9 Dictionaries* the manner of encoding lexicographic resources. The structure is a tree whose first nodes are annotated with the <entry> tag and treat every meaning and sub-meaning. TEI regulation:P5 covers a wide range regarding the annotation of lexicographic information, such as[2]:

– information about the form of the treated word (orthography, pronunciation, hyphenation, etc.);
– grammatical information (part of speech, grammatical sub-categorization, etc.);
– definitions or translations into another language;
– etymology;
– examples;
– usage information;
– cross-references to other entries;
– notes;
– entries (often of reduced form) for related words, typically called related entries.

---

[1] http://www.tei-c.org/Guidelines/P5/.

[2] The information was taken from the web-site http://www.tei-c.org/release/doc/tei-p5-doc/en/html/DI.html, subsection 9.2.2 Groups and Constituents.

3   Initiatives for Digitizing the Romanian Lexicographic Works

Dicţionarul limbii române (DA & DLR) is the most important lexicographic work edited by the Romanian Academy, which was started 110 years ago. The first edition was published in two series: DA (1913 – 1944, see [5]), DLR (1965 – 2010, see [6]) – 14 tomes, 36 volumes, over 20000 lexicon type pages and having between 7000 and 11000 characters on page. The richest and the most complex of the Romanian dictionaries, *Dicţionarul limbii române* is a thesaurus-type dictionary that is historical, explanatory, with normative, grammar, use and etymologic indications. 175000 lemmas are grammatically described, defined by glosses followed by total or partial synonyms, with the translation of the main meanings in the French language (only in DA), with examples that mention the source, extracted from a Bibliography that is formed of representative texts for the written and spoken Romanian language, since the first attested Romanian documents, at the beginning of the 16th century, until the current time. All the functional levels of the popular and literary language are taken into consideration. All the regionalisms and archaisms are registered. We have also included the neologisms that tend to enter the general use, by the attestation of their presence in texts that belong to different stylistic registers (scientific and publicistic, for instance). There are also indicated lexical variants, use-related information, internal correlations (between entries and meanings in the dictionary) and external correlations (by citing, for every meaning, dictionaries of the Romanian language from the Bibliography, in DLR). Each entry ends with information regarding pronunciation, writing, lexical variants and etymology.

The second edition of DLR was started in 2010 and will be completely elaborated in electronic form. This new project is called *DLRi* (*Dicţionarul limbii române informatizat – Romanian Language Computerized Dictionary*[3]) and was started with the electronic acquisition of the textual resources of the Bibliography (the Romanian language doesn't have an electronic diachronic corpus), using a collection of sheets, comprising of contexts for the letters A, C, and F, from manuscripts, texts written in Slavonic alphabet (sec. 17th–19th century), modern and contemporary texts, and it is written in an electronic editing interface, adapted by Oxygen (work in progress). The project which prepared the elaboration of the second (electronic) edition was eDTLR (***Dicţionarul tezaur al limbii române în format electronic** – The Romanian Thesaurus Dictionary in Electronic Format*, 2007–2010), whose main objective was the acquisition in electronic format of the complete form of DA & DLR (see [7,8]), as result of retro-digitalising[4]. eDTLR has a site of itself and can be searched through, by an independent interface. The project, partially financed, still has to go through

---

[3] Ongoing project of the Romanian Academy specialized Institutes.

[4] eDTLR. *Dicţionarul tezaur al limbii române în format electronic* (2007–2010) includes the old series of Dicţionarul limbii române (DA) and the new series of Dicţionarul limbii române (DLR), edited under the aegis of the Romanian Academy http://www.academiaromana.ro/.

a series of stages: building special parsers for the DA volumes that are formally non-homogeneous, refining the DLR parser, indexing the citations in the volumes of the DLR Bibliography (see [9]), final correction of the text that currently has errors due to the OCR process, validating the sense trees[5]. "The parser implemented a three steps process. In the first step a configuration of markers is used to identify the different types of fields in the entries. The sense tree of each entry is then determined by exploiting another level of markers and, finally, the atomic definitions (fine-grained senses) are extracted by means of a third level of markers. As such, the parser first identifies the sense markers in a breadth-first manner, and only afterwards builds the sense tree" (see [3]).

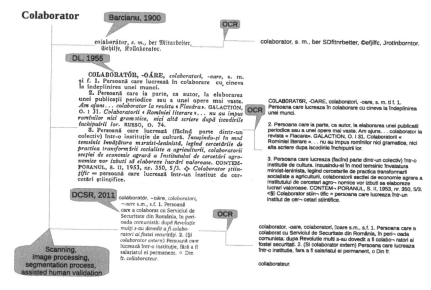

Fig. 1. CLRE - representation of title word "colaborator" (collaborator)

The project **CLRE. Corpus lexicografic românesc esențial. 100 de dicționare din bibliografia DLR aliniate la nivel de intrare**[6] (*Essential Romanian Lexicographic Corpus. 100 dictionaries from DLR Bibliography aligned by entries*) – national financing between 2010 and 2013 – is a natural continuation of the projects which dealt with the digitization of the DA DLR. *The objectives of the project* (see [10,11]) CLRE were: the accomplishment of a scanned corpus, with 100 dictionaries from the DLR Bibliography; scanning and processing of these dictionaries (by OCR – optical character recognition – the conversion from image to text; parsing the text at entry); elaborating

---

[5] http://metashare.info.uaic.ro/repository/browse/edtlr-the-thesaurus-dictionary-of-the-romanian-language/c05a74c063d611e28e8252540060617d5d02cd135a8b475 9904cb7a956d41aa7/.

[6] http://lexi.philippide.ro/clre/.

an online interface for validating/correcting the parsing information, validating the alignment between the text of the eDTLR and the dictionaries from the DLR bibliography. This project applies both classical/traditional linguistic methods (for example, transliterating the entries from the Slavonic or the transition Romanian alphabet and the comparative study of the dictionaries), as well as new, lexicographic-computational methods. From 2014, CLRE became a project of the Romanian Academy (Figs. 1, 2, 3, 4 and 5).

From de very beginning, the CLRE project was conceived as a linked database, because, on one hand, the title words from the CLRE list are related with dictionaries where they are treated in a lexicographical manner, and, on the other hand, there are links between CLRE list of words and eDLTR entries.

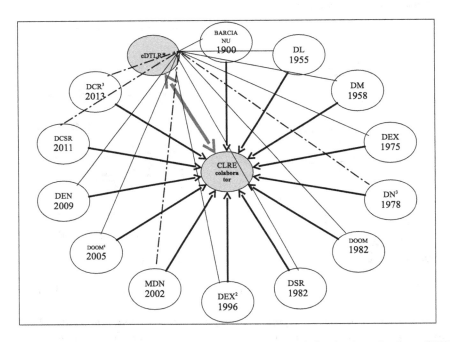

Fig. 2. The links between CLRE lemma "colaborator" and the dictionaries from CLRE corpus; the link between CLRE and eDTLR lemmas; the links between definitions of CLRE and those of eDTLR. (continuous line means same grammatical form, same meanings; discontinuous line = same grammatical form, new meaning(s).) *According to eDTLR, the first occurrence of word "collaborator" is about 1855.

There are some Romanian lexicographic resources available on-line within the Romanian language area. The first attempt which came close to the project CLRE, is dexonline[7], where a number of 44 dictionaries were textually published on the Internet and represented a starting point for the public interested

---

[7] http://dexonline.ro/.

in the Romanian language. It was developed and managed by volunteers. The definitions are manually inserted from the sources, except obvious typographical errors that were corrected. Access to dexonline is free. Some of the definitions of corpora can be downloaded for free under the GNU - General Public License. The particular difference that the project CLRE brings is represented by the vast number of dictionaries taken into account (more than 100 dictionaries from DLR Bibliography), adding up to approximately 150.000 dictionary pages, which have to be included in the database and aligned by entry level; there is also the historic perspective on the Romanian lexicography and, implicitly, on the Romanian language; the fact that what we propose is a corpus realized by all scientific norms, elaborated by a heterogeneous team, made of linguists and IT specialists and that the evaluation of results is made by Romanian renowned specialists, ensures the quality of a modern research.

Another Romanian lexicographic resource available on-line is *Lexiconul de la Buda – The Lexicon of Buda*, the electronic edition[8] of the first etymological and explanatory dictionary of the Romanian language, a benchmark for modern Romanian lexicography. For specialists and not only, which consults the dictionary online, the access is free after registration on the site by creating an account and password, the interface is easily accessible, designed to be functional, useful, permitting different query criteria, thereby allowing a quick rundown of the text (in four languages: Romanian, Latin, Hungarian and German). The current electronic edition restores the volume that was printed at the Buda press, in the 1825, under the complete title of *Lesicon romanescu-latinescu-ungurescu-nemtescu quare de mai mulţi autori, in cursul a trideci, si mai multoru ani s'au lucrat. Seu Lexicon valachico-latino-hungarico-germanicum quod a pluribus auctoribus decursu triginta et amplius annorum elaboratum est.* (LB, see [4])*Romanian-Latin-Hungarian-German Lexicon Elaborated by Various Authors Over More Than Thirty Years.*

The authors of this work specify on the project web-site: "Being accessible to the specialists and general public alike, the outcome of this project allows not only a faster access to the text, but a complex examination, as well. The results of these different types of search (information on etymons, on the status of a word - inherited or borrowed -, on internal creations, on word correspondence in various languages) are randomly displayed, leaving it to the specialists to decide on one value or another" (further information can be found on the site of this resource).

4   Types of Dictionaries. CLRE, DLR

CLRE digitizes dictionaries from the DLR Bibliography, which contains the quasi total number of Romanian dictionaries elaborated during the 17th–19th centuries and in the first half of the 20th century. For the second half of the past century and the period after 2000, we operated a value selection. Currently, the DLR Bibliography includes 391 lexicographic works, out of which for now 118 volumes were digitized in electronic format in CLRE, another 60 being in various stages of processing.

---

[8] http://www.bcucluj.ro/lexiconuldelabuda/site/login.php.

The typological diversity of these lexicographic works is large, and the nature and manner of presentation of the lexical and semantic information is determined by this typology. We shall present, *infra*, the types of CLRE dictionaries taking into consideration a part of the criteria[9] for indexing data for the dictionaries from the ENeL[10] corpus.

All the dictionaries are accessible on-line, through personalized accounts, permitting searches at lemma level, through the CLRE web-site. The access to these dictionaries is not public, only the specialists can use the corpus, for research purposes.

The CLRE corpus is lined at lemma level with *Dicționarul limbii române* in electronic format[11] (work in progress) and with e-TIKTIN (H. Tiktin, Rumänisch-deutsches Wörterbuch, 3rd edition, TDRG[3], see [4] - in electronic format)[12]. This is the second Romanian dictionary fully digitized in XML format, based on parsing heuristics.

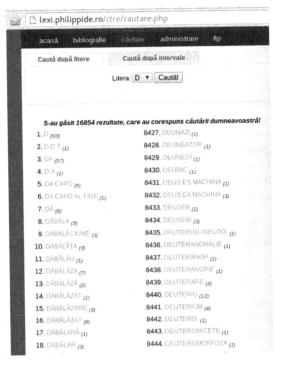

Fig. 3. CLRE - The list of words with the letter D in dictionaries validated within CLRE project, indicating the number of terms and the number of dictionaries defining each term.

The dictionaries that are already included in CLRE can be grouped by the following criteria (the examples do not limit the list of dictionaries and all of them can be seen in [4]):

---

[9] Language. Accessibility (not online, partly online, online, pay wall partly accessible). Digitization and mark-up language (HTML full text digitization, page image scan with OCR, page image scan). Search on (lemma, lemma or meaning, not possible). Form (audio, no audio). Grammatical Characterization (Y/N). Meaning/definition (no definition, synonyms, paraphrases, translation, image). Etymology (Y/N). Examples (no, with source, without source). Usage information (Y/N). Cross-references (no, other entries, other dictionaries). Dictionary Portal (Y/N). Metalanguage. Etymological (Y/N). Dialectal information (Y/N).

[10] http://www.elexicography.eu/.

[11] Within CLRE, a part of the DA and DLR volumes were rescanned, fully reparsed at entry level and validated at lemma level.The CLRE aligning is made with this form of DA+DLR.

[12] www.lexi.philippide.ro/tiktin.

TIME OF CREATION: **old** dictionaries (1650–1830), **modern** dictionaries (1831–1946), **contemporary** dictionaries (1947–present). The old and modern dictionaries are edited in many alphabets: Latin, old Bulgarian or old Slavonic, Cyrillic, gothic German, in the transition alphabet from the Cyrillic alphabet to the Latin alphabet (1840–1862).

LANGUAGE: **monolingual** (I.D. Negulici, *Vocabular român de toate vorbele străbune...*, NEGULICI, see [4] – *Romanian Vocabulary of all ancient words* ...; Lazăr Șăineanu, *Dicționar universal al limbei române*, ȘĂINEANU, D.U., see [4] – *Universal Dictionary of the Romanian Language...*; *Dicționarul limbii române literare contemporane*, DL, see [4] – *The Dictionary of the Contemporary Literary Romanian Language*; *Micul dicționar academic*, MDA, see [4] – *The Small Academic Dictionary*), **bilingual** (*Dictionarium valachico-latinum*, see [4]; Andrei Iser, *Vocabular românesc-nemțesc*, ISER, see [4] – *Romanian-German Vocabulary*; Frédéric Damé, *Nouveau dictionnaire roumain-français...*, DAMÉ, see [4]; H. Tiktin,

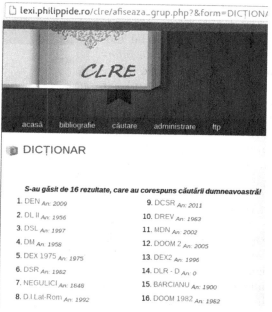

Fig. 4. CLRE - Dictionaries list that define the word "dictionary". Each dictionary has mentioned its year of printing.

Rumänisch-deutsches Wörterbuch), TDRG, see [4], **polyglot** (*Dicționariu rumânesc, latinesc și unguresc*, DRLU, see [4] – *Romanian-Latin-Hungarian Dictionary*).

LEXICOLOGICAL NATURE OF THE LEMMAS: **general dictionaries** (ȘĂINEANU, D.U.; *Dicționarul limbii române moderne*, DM, see [4] – *The Modern Romanian Language Dictionary*; *Dicționarul explicativ al limbii române* DEX, see [4] – *The Explanatory Dictionary of the Romanian Language*, **enciclopaedias** (Valer Butură, *Enciclopedie de etnobotanică românească*, BUTURĂ, E.B., see [4] – *Encyclopaedia of Romanian Ethnobotanic*; *Mic dicționar enciclopedic*, M. D. ENC., see [4] – *A Small Encyclopedic Dictionary*), **special dictionaries** (*Lexiconul tehnic român*, LTR, see [4] – *The Technical Romanian Lexicon*; BUTURĂ, E.B.).

MANNER OF DEFINITION: by **definition** (the majority), by **synonyms** (Mircea Seche, Luiza Seche, *Dicționar de sinonime al limbii române*, DSR, see [4] – *The Romanian Language Dictionary of Synonyms*), by **translation** (*Dictionarium valachico-latinum*), by definitions with **illustrations** (LTR).

PRESENCE OF GRAMMAR INDICATIONS: most of the dictionaries published after 1870 present grammar indications; some of the dictionaries are especially focused on the grammar specifications (*Dicționarul ortografic, ortoepic și morfologic al limbii române*, DOOM, see [4] – *The Orthographic, Orthoepic and Morphological Dictionary of Romanian Language*).

PRESENCE OF USE INFORMATION: this information can be **implicit**, for special dictionaries (*Dicționar de informatică*, D. INF., see [4] – *Informatics Dictionary*) or **explicit** (within the definition or by special indication) (ȘĂINEANU, D.U.); **presence of examples** (by mentioning the source): (Tamás Lajos, *Etymologisch-historisches Wörterbuch*, TAMÁS, ET. WB., see [4]).

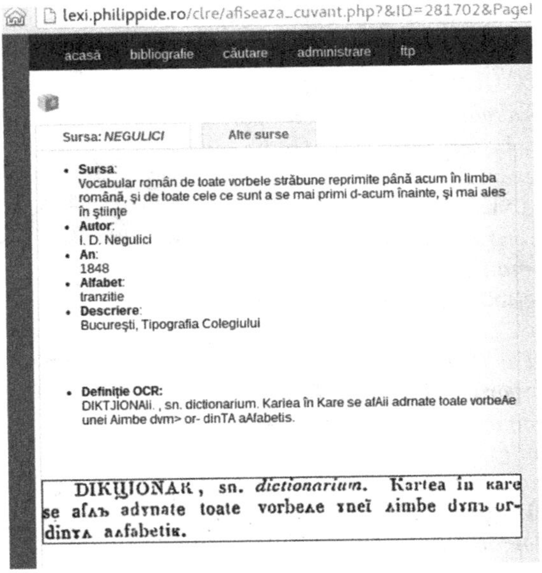

PRESENCE OF ETYMOLOGICAL INFORMATION: the general and encyclopaedic dictionaries, but also the special dictionaries, such as those that analyse archaisms and neologisms, usually present etymological information, too; some special dictionaries treat the etymological explanations exclusively (among the first type, we find: (DM); among the etymological dictionaries, we mention: A. de Cihac, *Dictionnaire d'étymologie daco-romane*, CIHAC, see [4];

Fig. 5. CLRE - The term dictionary present in the work of I. D. Negulici, *Vocabular român de toate vorbele străbune reprimite pînă acum în limba română, și de toate cele ce sînt a se mai primi d-acum înainte, și mai ales în științe* [Romanian Vocabulary of all ancient words, readmitted so far in Romanian, and of all that are to be to received henceforth, especially in science], 1848, in which it uses a transition alphabet from Cyrillic writing to the Latin characters.

Alexandru Ciorănescu, *Dicționarul etimologic al limbii române*, CIORĂNESCU, D.ET., see [4] – *The Etymological Dictionary of Romanian Language*).

PRESENCE OF DIALECTAL INFORMATION: beside general dictionaries (DM) that also include this type of information, we also find a series of special dictionaries and glossaries of regional terms, such as: *Dicționarul limbii române literare vechi (1640–1780). Termeni regionali*, DLRV, see [4] – *The Old Literary Romanian Language Dictionary (1640–1780). Regional terms*; *Glosar de cuvinte dialectale din graiul viu al poporului român din Ardeal*. Adunate și explicate de Alexiu Viciu, VICIU, GL., see [4] – *Glossary of Dialectal Words of the Living*

Speech of the Romanian People from Transylvania. Gathered and explained by Alexiu Viciu; *Glosar dialectal,* COMAN, GL., see [4] – *Dialectal Glossary.*

5 COST – ENeL

In 2013 it started a new project called ENeL: *European Network of e-Lexico-graphy*[13], a COST Action. This project aims at enlighten the European lexicography and points out the common pan-European linguistic heritage. Even if there are examples of research in the field of e-lexicography in Europe, there is a lack of common approaches and of common standards. An European Network of e-Lexicography would enable the academic lexicographical community to co-ordinate and harmonise research in the field of e-lexicography. The COST network aims at establishing common standards with respect to technologies for both retro-digitized and new electronic dictionaries and at developing a join approach to e-lexicography and new methods in lexicography that fully explore the possibilities of the digital medium and reflect the pan-European nature of much of the vocabularies of the languages of Europe. Common standards and joint approaches are beneficial to all dictionary projects, particularly to those documenting small languages, which often do not have the capacities to develop their own standards.

As mentioned in the Memorandum, the Aim of this Action is "to increase, co-ordinate and harmonise European research in the field of e-lexicography and to make authoritative dictionary information on the languages of Europe easily accessible. The Action will (a) make lexical knowledge of small and large languages available in an European dictionary portal; this portal will serve as the central reference point for all dictionary users who look for reliable, authoritative dictionary information on the languages of Europe and their histories on the Internet. The Action will (b) enable cooperation and exchange of resources, technologies and experience in e-lexicography and provide support for dictionaries which are not yet online. It will (c) discuss and aim at establishing standards for innovative e-dictionaries that fully exploit the possibilities of the digital medium. In taking a more common approach to e-lexicography the Action will (d) establish new ways of representing the common heritage of European languages by developing shared editorial practices and by interconnecting already existing information"[14].

By creating an European dictionary portal with information on and access to the academic dictionaries of Europe the Action will make authoritative dictionary information visible for a much wider audience and easily accessible to other lexicographers, researchers from other disciplines, writers and translators and to the general public. For small language dictionaries such a portal is an opportunity to present themselves without worrying about printing costs – thus the portal is also a means to preserve the European language diversity. The large amounts of data connected in the European dictionary portal will enable

---

[13] http://www.elexicography.eu/.

[14] http://www.cost.eu/COST_Actions/isch/Actions/IS1305.

new lines of research in the field of digital humanities that could not have been carried out on the basis of isolated language resources, e.g. tracing the spread of technological innovation by studying their appearance in the vocabularies of the languages of Europe.

Through the four *Working Groups of this Action* (WG 1 *Integrated interface to European dictionary content*, WG 2 *Retro-digitized dictionaries*, WG 3: *Innovative e-dictionaries* and WG 4 *Lexicography and lexicology from a pan-European perspective*) the Action will coordinate and increase research in the field of (electronic) lexicography across Europe through the establishment and progressive growth of an European Network of e-Lexicography.

In order to create an European dictionary portal, the members of the Action gathered information about scholarly dictionaries from their countries. Until the middle of 2015 the list had about 240 titles and the purpose is to make authoritative dictionary information on the languages of Europe accessible to both the general public and to lexicographers in an European dictionary portal. The portal will list and give information on scholarly dictionaries of the languages of Europe and provide access to these dictionaries. It will allow sorting by 22 parameters already mentioned above.

The large amounts of data connected in the European dictionary portal will enable new lines of research in the field of digital humanities that could not have been carried out on the basis of isolated language resources, e.g. tracing the spread of technological innovation by studying their appearance in the vocabularies of the languages of Europe.

6   Perspèctives of Correlation and Exploitation of CLRE Lexicographic Linked Data

The effort made within CLRE was to put into digital format a series of lexicographic works of great importance for the Romanian language. The work done until now (CLRE is an initiative to which we actively work) has a tremendous size; nevertheless, the digitally annotated lexicographic information is poor. At this moment, within the database, each processed dictionary (scanning, graphic processing, OCR, indexing, validating the segmentation step) contains for each identified entry the corrected title-word, an OCR recognition of its definition field and a image capture of the place in the printed page that contains the mentioned elements.

As we know, the OCR process is not viable in terms of accuracy of words recognition and, for this reason, noises are inserted in the texts operated through this process. These recognition problems are caused mainly by the quality of the printed material, by the digital acquisition conditions and mechanisms of the scans. Thus, most of the times, the texts obtained in this manner cannot represent a starting-point for further linguistic-computer processing. For this reason, it is necessary to perform a correction of the text content. The development group is analysing the possibilities mentioned below.

Proof reading by human intervention would considerably improve the corpus quality, but it is impracticable if we consider the long period of time necessary to work in this direction. Moreover, we mention that the development team is not numerous[15] and, scientifically speaking, this work method does not offer important satisfaction.

The alternative would be the automatic correction. Unfortunately, for the Romanian language there are no instruments available, necessary for accomplishing this task with notable results. The proof reading of the lexicographic text is a difficult task because of the combination between the lexicographic language and the common language in the same text.

A solution analysed by the development group is the comparison between the texts of the definitions obtained within CLRE and those obtained in similar projects. eDTLR, dexonline are main references that could be used for this purpose. Moreover, the development of Romanian language corpora is watched with interest, because probabilistic methods based on n-grams can be used.

In order to develop such a project, it is necessary to create connections with online existing digital resources. For instance, we take into consideration the creation of links with linguistic atlases for Romanian language, which also include extremely valuable lexical information (actually recovered in the DLR). Thus, starting from the last two volumes of *Noul Atlas lingvistic român pe regiuni. Moldova şi Bucovina* (NALR -MB, see [4]) (*The new Romanian linguistic atlas on regions. Moldavia and Bucovina*), edited with the help of programs for electronic writing and map projection, direct links could be made in eDTLR. A necessity for creating connections with the atlases is the digitization the other similar works of Romanian linguistic geography (published on paper during the past century). Moreover, connections could be made with the information from the Wikipedia-type encyclopedic platforms, operation that would determine the increase of visibility of lexicographic data included in eDTLR/CLRE, but also the enrichment of information to which the reader/user has access.

The access to the information from dictionaries can be adapted and exploited in various other applications, also by including the vocal/audio variant of dictionaries (in the electronic form of the volumes, the pronunciation of the title-word can be included by default).

Last, but not last, with the help of bilingual dictionaries, this information can be correlated with resources of other languages, especially European languages, offering the public research materials useful both for knowing the Romanian language and also for aligning the Romanian lexicography to the contemporary trends in the field (see [12–14]).

---

[15] After finalising the financing contract, the CLRE project became a priority project of the Romanian Academy. The development team was formed of 4 persons, members of the Lexicology-Lexicography Department of the Institute of Romanian Philology within the Romanian Academy. In addition to the four members, the project has enjoyed the contribution of a small group of volunteers.

References

1. Dictionarium valachico-latinum. Primul dicționar al limbii române, Studiu introductiv, ediție, indici și glosar de Gh. Chivu. Editura Academiei Române, București (2008)
2. Cristea, D., Radu, E., Moruz, A., Răschip, M., Patrașcu, M. I.,Tufiș, D., Haja, G., Teodorescu, H.I., Curteanu, N., Munteanu, E., Marian, R., Busuioc, M.: Scientific and Technical Report: eDTLR – Dicționarul Tezaur al Limbii Române în format electronic, December 2010
3. Cristea, D., Haja, G.: The Thesaurus Dictionary of Romanian Language in Electronic Form, in "Clarin", Newsletter of Clarin Project, no. 13, pp. 10–11, January-June 2011. http://www.clarin.eu.newsletter
4. Bibliografie. În Dicționarul limbii române (DLR). Serie nouă. Redactori responsabili: acad. Marius Sala și acad. Gheorghe Mihăilă. Tomul I, Partea a 3-a. Litera D (D – deînmulțit), pp. XIX–XCV. Editura Academiei Române, București (2006)
5. Dicționarul limbii române (DA). Sub conducerea lui Sextil Pușcariu. Tomul I. Partea I: A – B. București, Librăriile Socec & Comp. și C. Sfetea (1913); Tomul I. Partea II: C. București, Tipografia Ziarului "Universul" (1940); Tomul I. Partea III. Fascicula I: D– de. Universul, Întreprindere Industrială a Statului, București (1949); Tomul II. Partea I: F – I. Monitorul Oficial și Imprimeriile Statului. Imprimeria Națională, București (1934); Tomul II. Partea II. Fascicula I: J – lacustru. Tipografia Ziarului "Universul" S. A., București (1937); Tomul II. Partea II. Fascicula II: Ladă – lepăda. Tipografia Ziarului "Universul" S. A., București (1940); Tomul II. Partea II. Fascicula III: Lepăda – lojniță. Tipografia Ziarului "Universul" S. A., București (1948)
6. Dicționarul limbii române (DLR). Serie nouă. Redactori responsabili: acad. Iorgu Iordan, acad. Alexandru Graurși acad. Ion Coteanu. Din anul 2000, redactori responsabili: acad. Marius Sala și acad. Gheorghe Mihăilă. Editura Academiei Române, București, Tomul I. Partea a3a, a 4a, a 5a, a 6a: Litera D (2006–2008); Tomul IV. Litera L (2008); Tomul V. Litera L (2008); Tomul VI. Litera M (1965–1968); Tomul VII. Partea I. Litera N (1971); Tomul VII. Partea a 2a. Litera O (1969); Tomul VIII. Litera P (1972–1984); Tomul IX. Litera R (1975); Tomul X. Litera S (1986–1994); Tomul XI. Partea 1. Litera Ș (1978); Tomul XI. Partea a 2a și a 3a. Litera T (1982–1983); Tomul XII. Partea I. Litera Ț (1994); Tomul XII. Partea a 2a.Litera U (2002); Tomul XIII. Partea I și a 2a și a3a. Litera V și literele W, X, Y (1997–2005); Tomul XIV. Litera Z (2000)
7. Haja, G., Dănilă, E., Forăscu, C., Aldea, B.-M.: Dicționarul limbii române (DLR) în formatelectronic. Studii privind achiziționarea. Editura Alfa, Iași (2005)
8. Cristea, D., Răschip, M., Forăscu, C., Haja, G., Florescu, C., Dănilă, E., Aldea, B.: The digital form of thesaurus dictionary of the Romanian language. In: Burileanu, C., Teodorescu, H.-N. (eds.) Advanced in Spoken Language Technology, pp. 195–206. Editura Academiei Române, București (2007)
9. Cristea, D., Răschip, M.: Linking a digital dictionary ontoits sources. In: Proceedings of the FASSBL 2008, Dubrovnik, Croatia, September 2008
10. Tamba Dănilă, E., Clim, M.-R., Catană-Spenchiu, A., Patrașcu, M.: The evolution of the romanian digitalized lexicography. The essential romanian lexicographic corpus. In: Fjeld, R.V., Torjusen, J.M. (eds.) Proceedings of the 15th EURALEX International Congress, 7–11 August 2012, pp. 1014–1017. Press Reprosentrales, UiO (2012). http://www.euralex.org/proceedings-toc/euralex_2012

11. Tamba, E.: La lexicografía Rumana. Historia y Aactualidad. In: Rodríguez, F.C., Seoane, E.G., Palomino, M.D.S. (eds.) Lexicografía de las lenguas románicas. Perspectiva histórica, vol. I, pp. 265–282. de Gruyter Verlag (2014)
12. Tamba, E., Clim, M.-R., Catană-Spenchiu, A., Patrașcu, M.I.: Situația lexicografiei românești în context European. Philologica Jassyensia **VIII**(2(16)), 259–268 (2012). http://www.philologica-jassyensia.ro/upload/VIII_2_Tamba_Clim.pdf
13. Dănilă, E.: Corpus lexicographique roumain essentiel. Les dictionnaires de la langue roumaine alignés au niveau del'entrée. In: Herrero, C., Rigual, C.C. (eds.) Actas del XXVI Congreso Internacional de Lingüísticay de Filología Románicas, Valencia, 6–11 Septiembre 2010, vol. VIII, pp. 125–134. de Gruyter Verlag (2013)
14. Clim, M.-R.: La lexicografía rumana informatizada: tendencias, obstáculos y logros. In: Sánchez Palomino, M.D., Domínguez Vázquez, M.J. (coords.), Domínguez Vázquez, M.J., Guinovarty, X.G., Riveiro, C.V. (eds.) Lexicografia de las lenguas románicas. Aproximaciones a la lexicografía moderna y contrastiva, vol. II, pp. 95–110. de Gruyter Verlag (2015)

Sentiment Analysis in Social Media
and Linked Data

Extracting Features from Social Media
Networks Using Semantics

Marius Cioca[1], Cosmin Cioranu[2], and Radu Adrian Ciora[1(✉)]

[1] Faculty of Engineering, "Lucian Blaga" University of Sibiu, Sibiu, Romania
{marius.cioca,radu.ciora}@ulbsibiu.ro
[2] Executive Unit for Higher Education, Research,
Development and Innovation Funding, Bucharest, Romania
cosmin.cioranu@uefiscdi.ro

Abstract. This paper focuses on the analysis of social media content generated by social networks (e.g. Twitter) in order to extract semantic features. By using text categorization to sort text feeds into categories of similar feeds, it has been proved to reduce the overhead that is required to retrieve these feeds and at the same time, it provides smaller pools in which further investigations can be made easier. The aim of this survey is to draw a user profile, by analysing his or her tweets. In this early stage of research, being a pre-processing phase, a dictionary based approach is considered. Moreover, the paper describes an algorithm used in analysing the text and its preliminary results. This paper is focusing to support research in Social Media exploration. Thus, it describes a tool useful for communication experts to analyse public speeches. So far, this tool gave promising results in inferring socio-political trends from social media content of public speakers. We also evaluated our experiment on Support Vector Machine (SVM) with 10-fold cross-validations.

Keywords: Public speech · Feature extraction · Social media · Semantic analysis · Support Vector Machine · Linking online data

1 Introduction

Due to the large volume of data that has gradually increased in all areas, there is a constant need to develop new methods, techniques and solutions necessary for harnessing the power of this information stored in this data. Useful information is almost impossible to extract from such large amounts of data, with traditional methods based on human analysis capacity.

In this context, the paper presents, in a data mining process, the pre-processing step for the analysis of the text and the extraction of features from Social Media (e.g. Twitter) with the aim of creating user profiles in order to determine users orientation towards certain predefined patterns. Thus, a solution is presented regarding the unstructured data pre-processing, as well as proposals concerning the semantic integration of the data imported from different information sources.

© Springer International Publishing Switzerland 2016
D. Trandabăţ and D. Gîfu (Eds.): EUROLAN 2015, CCIS 588, pp. 127–136, 2016.
DOI: 10.1007/978-3-319-32942-0_9

The solution regarding the extraction of knowledge has focused on unstructured data sources, namely written text in natural language extracted directly from internet, more specifically – Twitter feeds.

Supervised machine learning In order to validate our initial results, a new approach to text categorization was designed, by using learning algorithms. From these algorithms, support vector machines proved to be the most suitable for text categorization. As this is a pure theoretical approach, it does not require any parameter adjustment.

By using text categorization to sort text feeds into categories of similar feeds, it has been proved to reduce the overhead that is required to retrieve these feeds and at the same time, it provides smaller pools in which further investigations can be made more easily.

The paper is structured as follows: Sect. 2 describes a brief previous work, offering a picture of contributions in the extracting the semantic features. Section 3 describes the methods that were used for data analysis. In Sect. 4, the algorithm that we propose is detailed and the use of support vector machines is explained. The following Sect. 5, briefly presents the achievements. Finally, Sect. 6 presents the conclusions and several challenges for the future work.

2   Previous Work

In the early of 20th century, the basis of quantitative content analysis had been laid in USA [10]. For the analysis of the public speeches, sets of words are defined, which fall into specific categories, like for example, "allusion to the common past" includes expressions like: "our glorious past" or "our memories". In the quantitative analysis, the pre-processing phase has a very important role, as it selects the relevant features that are sought, thus eliminating the so called noise which comprises of link-words and other elements that are considered irrelevant to the problem that is sought to be solved.

In [9], Joachims explores the use of Support Vector Machines (SVMs) for training text classifiers using examples. The paper analyses the specific properties of learning from text content and determines the reasons why SVMs perform well in these tasks. The theoretical analysis of this paper concludes that SVMs acknowledge the following particularities of text documents: high dimensional feature spaces, few irrelevant features and sparse instance vectors. The paper concludes that SVMs consistently achieve constant good performance. It also finds out that they generalize well in high dimensional feature spaces, thus eliminating the need for manual feature selection and so netting the way of its usage in text categorization. It is also stated that SVMS are robust compared to traditional methods, thus having a good performance in experiments, ceasing to fail where conventional methods do. Moreover, SVMs do not require parameter tuning, as they accord themselves automatically.

In [4], Crammer and Singer describe an algorithm implementation of a multiclass kernel-based support vector machines. Their approach was based on the margin notion of multiclass problems. Thus, using this notion, they transform the problem of categorization into a constrained optimization problem defined by a quadratic objective function. Their use of the dual of the optimization problem enabled them to use kernels with a small set of constraints and split the dual problem into multiple problems of smaller size. They also describe an efficient algorithm for solving the reduced optimization problem and also show that is convergent.

Comparing with other works, on empirical text pre-processing [5–7] we propose a different interpretation of results. In addition, the results are presented directly to the user. Actually, this method is inspired from another survey, presented in [8], in which the idea was to integrate a wide range of language processing tools, in order to build a complex characterisation of the political discourse. The platform developed in [8] helps to outline distinctive features which put discursive characteristics of political figures into a new and sometimes unexpected light.

Our approach can process tweeter feeds in almost real-time, comparing with the analysis of static text bodies, found it in the study mentioned. The use of SVMs has made possible the seamless extension of our categories without being necessary to do any alteration to existing algorithm.

The development of a toolkit able to transform the data from an original format into triples that would be stored in a triple store would make easier the interpretation by any computer or application that processes these data [1, 2].

3   Work Method

The method used for analysing the various data sources aims to establish the user profile, by extracting features based on a dictionary composed of classes that contain words.

In order to validate our initial results, a new approach to text categorization was designed, by using learning algorithms. From these algorithms support vector machines proved to be the most suitable for text categorization. As this is a pure theoretical approach, it does not require any parameter adjustment.

For applying the method, by means of the PHP (Hypertext Pre-processor) language, a software application was designed for retrieving, classifying and interpreting the information from various sources, offering the possibility of analysing the data both from the text files and the data available on the Internet (e.g. Twitter, RSS, etc.) [3].

In this survey, the data of certain political leaders was analysed, extracted from the addresses found online on Twitter, tracing the occurrence frequency of various words belonging to certain classes that make up the dictionary. The vocabulary used in the application consists of a total of 30 classes (swear, social, family, friends, people, emotional, positive, negative, anxiety, anger, sadness, rational, intuition, determine, uncertain, certain, inhibition, perceptive, see, hear, feel, sexual, work, achievements, failures, leisure, home, financial, religion, nationalism), with each class comprising of a total of 8–10 words.

The class name is represented by the name of the .txt file, which contains the words of the class. The dictionary also contains a *special class* made of *linking words*, which are used to eliminate them from the analysed text. Therefore, as these are text files, anyone can add as many words in a class or can add as many classes they want to the dictionary. The more elaborated the dictionary is, the better the accuracy of the results.

An example of class used in the dictionary is the *Social Class*, represented by a text file (social.txt) with the following content:

```
name = 'social'
w [] = help
w [] = celebration
w [] = consultation
w [] = event
w [] = intervention
w [] = interview
w [] = communication
w [] = friendly
w [] = sociable
w [] = voluble
```

The links that are to be analysed are added also in a text file (link.txt), their limit being determined only by the processing power of the system used. For the results presented in the next chapter, the analysed data (text written in natural language) comes from the following addresses:

```
l[]="0::https://twitter.com/barackobama"
l[]="0::https://twitter.com/angela_d_merkel"
l[]="0::https://twitter.com/david_cameron"
l[]="0::https://twitter.com/putinrf_eng"
```

Because all defined and used classes (categories) by the application are balanced from a quantitative point of view (they contain ten words), in the application only the qualitative normalization process is performed. The application also uses a special class that contains connection words; those were eliminated, in order to reduce the noise of the results obtained in the data pre-processing phase.

Supervised machine learning tasks usually consist of labelling of instances, where labels are well defined by a finite set of elements. This is generally coined as multiclass learning. Usually, specialized algorithms were devised to multiclass problems, by building upon binary classification learning algorithms, i.e., problems in which just a couple of labels exist. Examples of this kind of algorithms are extensions of multiclass decision tree learning and boosting algorithms like AdaBoost.

4   Algorithm

In this section, we present the steps of the pre-processing algorithm used in analysing of the Twitter texts.

4.1   Pre-requirements

− The Base elements structure, which is basically, classes of words specified in a "source word" format − which are simple words without prefixes or suffixes.

– Links pool ready for classification

Note: These sources are similar to a standard configuration file, similar to an INI file, which is a simple text file with a basic structure composed of sections, properties and values.

4.2 Proposed Method

– Categories are loaded in an predefined format:

$$\texttt{wordCount}_{\texttt{cat}} = [\text{total number of words/category}]$$

– Every single link is downloaded.

   • The algorithm uses a source file, which contains web links and it crawls these links. It does not follow sub links, thus we say it operates at level 0.
– HTML Tags are eliminated and then a tree like structure is created, having the word as a root, and links go from it to all categories that this word is linked to.

   • Additionally there is a validation step which checks that each word has links to at least two categories.
– Words from the pool are looked for in the pages that the links point to and a distribution network of the words of form:

$$[\texttt{Word}(\texttt{root})]_{\texttt{cat/link}} = [\text{number of appearances in the linked web-page}]$$

The absolute probability is calculated from the total number of words and the number of their appearances into a given category:

$$PA_{cat/link} = \frac{\sum word(root)_{cat/link}}{wordCountTotal_{link}} \tag{1}$$

This absolute probability gives us an idea of how much the analysed discourse falls into the categories that we have defined.

– Normalized probability is then calculated:

$$PAFinal_{cat/link} = \frac{PA_{cat/link}}{\sum PA_{cat/link}} 100. \tag{2}$$

The normalized value of the probability gives an overall interpretation of the analysed discourse as a whole, trying to infer its trend.

The need for Multi-class Support Vector Machine was determined by the necessity of empowering the user of our tool to the increase the total number of classes. As such, a Multi-class SVM algorithm was implemented in order to be able to handle more and more classes. As any learning algorithms, it needed training at first. Ten of the classes were used for training and another ten were used for testing. We also tested the ability of learning of the algorithm with the rest of ten classes and as expected the algorithm automatically learned them and correctly categorized the input text that it was fed to it. We did not expect the machine learning to perform better in text categorization problem as see in the results, but rather in terms of processing speed.

SVMs are inherently two-class classifiers. The most common technique in practice has been to build a $|C|$ "one-versus-all" or OVA classification, and to choose the class which classifies the test datum with greatest margin. Another strategy is to build a set of one-versus-one classifiers, and to choose the class that is selected by the most classifiers. This involves building $|C|$ ($|C|$t–1)/2 classifiers, but the time for training classifiers may actually decrease, since the training data set for each classifier is much smaller.

However, these are not very elegant approaches to solving multiclass problems. A better alternative is provided by the construction of multiclass SVMs, where we build a two-class classifier over a feature vector $\Phi(\vec{x}, y)$ derived from the pair consisting of the input features and the class of the datum. At test time, the classifier chooses the class

$$y = \text{argmax}_y \vec{w}^T \Phi(\vec{x}, y')$$

The margin during training is the gap between this value for the correct class and for the nearest other class, and so the quadratic program formulation will require that

$$\forall i \forall y \neq y_i \vec{w}^T \Phi(\vec{x}_i, y_i) - \vec{w}^T \Phi(\vec{x}_i, y) \geq 1 - \xi_i$$

Classification for classes that are not mutually exclusive is called multilabel or multivalue classification. In this case, a document can belong to several classes simultaneously, or to a single class, or to none of the classes. A decision on one class leaves all options open for the others. It is sometimes said that the classes are independent of each other, but this is misleading since the classes are rarely statistically independent.

The algorithm for solving this classification task with linear classifiers is straightforward:

- The first step consists of building a classifier for each class, where the training set consists of the set of documents in the class (positive labels) and its complement (negative labels).
- Given the test document, apply each classifier separately. The decision of one classifier has no influence on the decisions of the other classifiers.

In the training phase of each class we used the elements of the class as positive labels and the elements of the other classes as negative labels, in order to ensure that no class overlapping will be possible. Thus in order for the algorithm to function properly, elements must exclusively belong to a single class.

5 Results

The achieved results - even in the pre-processing phase - are promising and heavily dependent on the accuracy and relevance of the words that compose the classes, which form the dictionary. It has been observed that incorrect training data that is fed into the SVM makes it less accurate in correctly categorizing the test data.

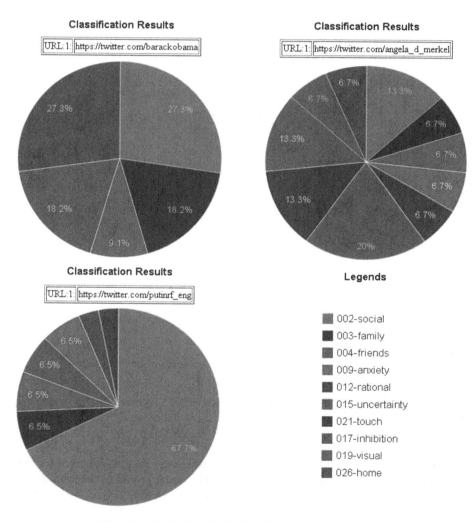

Fig. 1. Individual analysis of online resources (Twitter).

Table 1. Categorisation of online resources

Category	https://twitter.com/ barackobama	https://twitter.com/ angela_d_merkel	https://twitter.com/ putinrf_eng
002-social	–	13.3 %	67.7 %
003-family	18.2 %	6.7 %	6.5 %
004-friends	9.1 %	6.7 %	6.5 %
009-anxiety	–	6.7 %	6.5 %
012-rational	–	6.7 %	3.1 %
015-uncertainty	27.3 %	20 %	–
017-inhibition	18.2 %	13.3 %	6.5 %
019-visual	–	6.7 %	–
021-touch	–	13.3 %	–
026-home	27.3 %	6.7 %	3.1 %

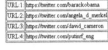

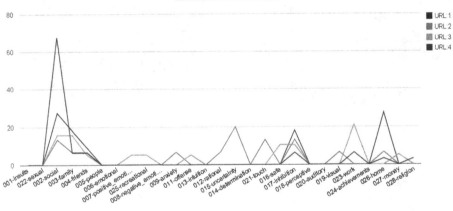

Fig. 2. A comparative analysis of the speeches' categorization (Twitter).

In Table 1, we present the results from 12/05/2015, where Twitter texts (e.g. Obama, Merkel and Putin) have been analysed using 10 semantic classes: social, family, anxiety, rational, uncertainty, inhibition, visual, touch, and home.

These are presented individually as in Fig. 1, where the information from a single link (online resource) is processed in order to establish the profile of the speaker.

The graphs were generated by integrating within the application the facilities offered by Google APIs.

The pie charts show that 67.7 % of Putin's speech has a strong social connotation. Also, more than half of Obama's speech is equally split between home and social topics.

From the analysis of Merkel's tweeter page, it can be very clear seen that her speech covers a wide range of topics with a small leaning towards uncertainty. A comparative analysis of the speeches' categorization is presented in (Fig. 2) where several links are analysed simultaneously and results are presented in parallel, in order to compare the results and profiles of these different tweets.

6 Conclusions and Future Work

The solution presented focuses on extracting the knowledge from unstructured data sources, namely text written in natural language. Even in this early pre-processing stage, it offers consistent results, which fuel us in deepening the research. The novel approach which makes use of multiclass support vector machines, provides an efficient learning algorithm and a practical implementation. The method used achieves state of the art results with a competitive running time.

Further on, the authors will focus on solutions based on the meta-data extracted from the data sources, semantic relations as well as lexical similarity algorithms to build a semantically annotated resource used within the semantic fusion algorithm.

In the future, we will focus on developing the current classes that contain words towards semantic classes with phrases and expressions, as well as elaborating certain meta-classifiers using algorithms, specific techniques and methods (Naive Bayes, Support Vector Machine, Neural Networks, etc.) simulated in MatLab and implemented in JAVA, on one hand.

References

1. Buraga, S.C., Cioca, M.: Using XML technologies for information integration within an e-Enterprise. In: The 7th International Conference on Development and Application Systems DAS, Under the Care of IEEE Romanian Section, Romania (2004)
2. Cioca, M., Buraga, S.C.: Using semantic web technologies to improve the design process in the context of virtual production systems. Int. J. Trans. Comput. 12 (2005)
3. Cioca, M., Ghete, A.-I., Cioca, L.I., Gifu, D.: Machine learning and creative methods used to classify customers in a CRM systems. In: DesPerrieres, O.D, Mazuru, S., Slatineanu, L. (eds.) Innovative Manufacturing Engineering. Applied Mechanics and Materials, vol. 371, pp. 769–773 (2013)
4. Crammer, K., Singer, Y.: On the algorithmic implementation of multiclass kernel-based vector machines. J. Mach. Learn. Res. 2, 265–292 (2002)
5. Gîfu, D., Cioca, M.: Online civic identity. Extraction of features. In: Soare, E. (ed.) Procedia – Social and Behavioral Sciences, vol. 76, pp. 366–371 (2013)
6. Gîfu, D., Cristea, D.: Computational techniques in political language processing: AnaDiP-2011. In: Park, J.J., Yang, L.T., Lee, C. (eds.) FutureTech 2011, Part II. CCIS, vol. 185, pp. 188–195. Springer, Heidelberg (2011)

7. Gîfu, D., Cristea, D.: Public discourse semantics. A method of anticipating economic crisis. Int. J. Comput. Commun. Control **7**(5), 832–839 (2012)

8. Gîfu, D., Cristea, D.: Multi-dimensional analysis of political language. In: Park, J.(J.H.), Leung, V.C.M., Wang, C.-L., Shon, T. (eds.) Future Information Technology, Application, and Service. Lecture Notes in Electrical Engineering, vol. 164, 1st edn, pp. 213–221. Springer, Heidelberg (2012)

9. Joachims, T.: Text categorization with support vector machines: learning with many relevant features. In: Nédellec, C., Rouveirol, C. (eds.) ECML 1998. LNCS, vol. 1398, pp. 137–142. Springer, Heidelberg (1998)

10. Lasswell, H.D.: Politics: Who Gets What, When, How. McGraw-Hill, New York (1936)

Social Data Mining to Create Structured Social Media Resources

Including Social Media – A Very Dynamic Style – in the Corpora for Processing Romanian Language

Cenel-Augusto Perez[1], Cătălina Mărănduc[1,2(✉)], and Radu Simionescu[1]

[1] Faculty of Computer Science, "Alexandru Ioan Cuza" University of Iaşi, Iaşi, Romania
{augusto.perez,catalina.maranduc,radu.simionescu}@info.uaic.ro,
catalinamaranduc@gmail.com
[2] "Iorgu Iordan - Al. Rosetti" Institute of Linguistics of the Romanian Academy,
Bucharest, Romania

Abstract. This paper aims to describe the process of introducing a new sub-corpus, in a new style, social media, in our UAIC-Ro-Dependency-Treebank. Our purpose is to enhance the corpus and to also include all the styles of the language. Unfortunately, the growth of the corpus is interrelated with the development of the syntactic parser. The inclusion of all the styles is a very difficult target; when parsing texts in a style for which the tools are not yet trained, the accuracy drops significantly. At least 1,000 sentences are needed for the first step of the training of the parser in a new style. We describe this first step that implies the introduction of social media style in the Treebank, the first series of orthographic, stylistic, pragmatic, lexical, semantic, syntactic, and discursive observations on this style of the language, and we communicate the first statistical evaluation of the automatic annotation.

Keywords: Treebank · Social media · Style of the language · Pragmatics · Stylistic analysis · POS-tagger · Syntactic parser · Rules formulation

1 Introduction

Social media is a means of communication that our contemporary society cannot do without. The conversations that take place in cyberspace satisfy the need for communication between people that are far away one from the other, and also between people that live closer but feel that this kind of conversation is more convenient. Despite the fact that this kind of communication is mostly in writing, on chat, on messenger or on blogs, it has many of the particularities of oral style because the people that communicate are usually in casual relationships or in socially equal relationships, and therefore the excessive politeness formulas and formalities are not common.

As two of the authors are linguists, they are interested in the study of stylistic peculiarities of this kind of communication. Textual analysis from a pragmatic and stylistic point of view will be facilitated by the existence of an annotated corpus that contains texts from social media communication. If we study this type of communication from a pragmatic point of view, we observe that the communication purpose is reached,

© Springer International Publishing Switzerland 2016
D. Trandabăţ and D. Gîfu (Eds.): EUROLAN 2015, CCIS 588, pp. 139–153, 2016.
DOI: 10.1007/978-3-319-32942-0_10

namely that the transmitter makes oneself understood by the addressees, though the form may be less rigorous or correct.

On the other side, the detection by the linguists of the pragmatic, lexical, stylistic peculiarities helps them identify the causes of the mistakes of the POS-tagger or of the syntactic parser and then to extract some rules in order to optimize the functioning of these tools, so as to increase the accuracy of their results into new parts of the corpus in the same stylistic variety.

UAIC-RoDepTreebank is already built respecting FDG axioms [14], of several sub-corpora, belonging of a certain stylistic variety. Therefore, we already experienced that whenever we introduce in the POS-tagger and then in the parser social media texts, belong to a stylistic variety on which the programs haven't been trained, the accuracy drops significantly.

We have two options. The first is to manually correct errors and in the same time to make comments over the causes that produce them, and to formulate rules that might remove disruptive factors. This method can be applied only for the POS-tagger, which supports the introduction of rules and restrictions. Our parser is not yet a mixed one, rule and statistic based. But after the observations on the supervised corpus, it would seem that wrong interpretations of the FDG parser are induced by the POS-tagging errors, hypothesis statistically confirmed. It would seem that by eliminating the POS-tagging errors, we will be able to also increase the accuracy of the FDG parser.

The second solution is the bootstrapping method to train the parser by corrected social media gold corpus. The two solutions don't exclude each other; therefore we will apply them both in different stages of our research.

2   The Structure of the Processed Social Media Sub-Corpus

We chose to select texts from chat or Messenger, because they are closer to the characteristics of oral language, familiar, being more spontaneous than blogs or comments. Because the privacy of people communicating via chat must be respected, we preferred to use of our own conversations. This has the consequence that the selected texts belong to people with a high cultural level.

It would be discussed which is the ratio of conversations between people with high cultural level and people with a precarious culture. Our hypothesis is that social strata are not isolated. So our corpus will contain at the same time shorthand expressions which have the communicative efficiency as purpose, and also the lexical and stylistic laboratory of the images designers (one of communicators is a writer and director of a literary journal, others are graduates of masters, PhDs, collaborators to literary journal).

This corpus is processed in .xml format and intended for advanced searches on computer. Therefore, the degree of discoverability is low. We also mention that we were not interested in the frequency of indecent terms, we have not specially selected them and we have never avoided if we met them. Should they be not used in mannerist form, like language tics without content, as it happens in certain social strata, they have a stylistic function, expressing the exasperation, frustration etc.

The Romanian lexicographers are prudish and they don't introduce familiar words for the sexual organs and taboo words although these words can be found in the dictionaries of other languages: therefore there are human and animal organs and functions which are not present in the dictionary, or are glossed with a scientific anatomic term, not with the familiar one. So, if an automatic machine translation which has incorporated such a dictionary, will translate a familiar term in the input language by a scientific term in the target language, the translation will be inadequate and ridiculous. (It is known that the inadequate selection of styles and their interference is the easiest way to obtain humorous effects). As the use of the indecent expressions is one of the familiar and chat language features, we will not hesitate to give some creative examples with the intention to lift the taboo.

The texts were copied from Messenger, and then have undergone a minimum preprocessing. We have removed very short notices repeated countless times, such as "thank you", "how are you", "bravo", etc., or we linked them to other replicas that were in the same relationship. We also put full stops at the end of all the sentences, because they are usually missing on chats, and without them the POS-tagger could not determine the boundary of each sentence.

We obtained 2,576 sentences, with 39,292 words and punctuations, by an average of 15.25 words per sentence. Of course, if the communicators had a lower degree of education the average should be lower. In our corpus, the short communications alternate with some quite long, usually of narrative type, written by persons which have the habit of composing texts.

We firstly processed this corpus with a POS-tagger automatic annotation, then with our FDG syntactic parser. We segmented this sub-corpus in two sections, with the intention of applying the bootstrapping method, and we checked the first bunch of 944 sentences, with 13,243 words and punctuation. In this paper, we will comment the results and we will perform a statistical comparison between the version obtained automatically and the corrected version.

The rest of our corpus, containing 16,136 sentences with 26,049 words and punctuation, will be checked after a second automatic annotation with the programs that we are improving after the observations during the check of the first part and we are trained on a gold corpus containing 944 chat sentences and a lot of texts with or without diacritics. Another series of observations during the supervision of its results will be exploited to the improvement of instruments, and so on.

3   Related Works

3.1   POS-Taggers for Social Media

Social media text are collected in corpora for different languages and they are applied for automatically detecting the users' opinion, for sentiment analysis, for study of the relationship between the members of a collectivity, for machine translations, information extracting and automatically resuming or for questions answering. The POS-tagging is the usual pre-processing step for all this applications. Of course, its quality influences subsequent processing performance. The POS tagger for the standardized language is achieving a high accuracy for each natural language. But social media texts are not standardized; like the familiar language, it is a very creative style, a laboratory for linguistic inventions, an empire of freedom that does not accept any regulations. So, the authors of all the papers describe methods to increase the accuracy of POS-taggers after an unsatisfactory start in the processing of social media texts.

In their papers [10–11], Neunerdt et al. describe a new corpus for German, composed of 36 000 annotated tokens, which contain web comments. Our corpus for Romanian is comparable, after three steps it is composed of 39,292 tokens which contain chat and Messenger dialogues. These corpora can be used for information extracting and also for training the POS-tagger. In these papers they describe a Marcov model POS tagger. Another model for the POS-tagger is a syntactic-semantic Bayesian HMM, described in [4]. These tools are generally trained for a particular language, as German, French, English, Chinese, Italian, Indonesian, etc.

Part of speech (POS) is one of the features that can be used to process data improving the quality of statistical-based machine translation. Typically, the language of POS is determined by the grammar of the language or is adopted from other languages. The work in aims to formulate a model to develop POS as linguistic factors for automatically improving the quality of machine learning. The method using a word similarity approach, perform clustering of the words contained in a corpus. Further classes will be defined as POS set obtained for a given language.

Many papers, such as [1, 3, 8], describe Twitter texts processing systems. In [13], the Frog tagger is combined with a post-processing module that incorporates the new, Twitter-specific tags in the Frog part-of-speech output. Running sequentially the Frog tagger and the post-processing module, it leads to a part-of-speech tagger for Dutch tweets. Approximately 1 million tweets collected in the context of the SoNaR project were tagged by Frog and the post-processor combined. Only a sub-set of annotated tweets have been manually checked. We should be inspired by this procedure to introduce the chat specific tags in the POS-tagger.

For increasing the accuracy of social media processing tools, there are two procedures: the normalization of the texts collected from the web [9] and from the blogs [5] or the inclusion of new modules and rules for the non-standardized texts in the POS-tagger. Procedures to adapt a POS-tagger for the noisy texts are described in [7, 17]. We are not interested in the first procedure, applied, as shown, only for marking the end of the sentences by period, in order to help the tool to detect it. We intend to create a corpus which will be used to study the particularities of web texts. If we check and remove its particularities and transform these non-standardized texts in standardized one, we cannot observe and comment them.

3.2 Syntactic Parsers for Social Media

The syntactic and semantic parsers for social media are less studied in the papers consulted. We intend to describe the syntactic particularities of this style of the communication and also to increase the parsing accuracy by increasing the gold corpus for the training. In [4, 6] we find a similar approach. The text processors, called parsers, are typically tested on news articles. Parsers can distinguish between words and punctuation, label parts of speech and analyze a sentence's grammatical structure. But these programs, in pipeline with the POS-taggers or not, don't do as well on social media. The authors of [3] wrote hundreds of rules to account for hash tags, repeated letters (as in "pleaaaaaase") and other linguistic features specific of Twitter. They also tried to use their program to distinguish between rhetorical questions and those that require a response. Businesses could use such a program to find what people are asking about their products. Their program classified correctly 68 % of 2 304 tweets.

4 Analysis of Social Media Style Particularities

4.1 Orthographic Particularities

A first observation is the fact that Romanian language orthography has specific letters with diacritic signs as: ș, ț, ă, î, â. The last two letters transcribed the same phoneme, the first appears at the beginning or at the end of a word, the second appears when the phoneme is not in an inner-word position. The user has to set the topic regional languages in order to be able to type these characters. Many users do not know the procedure or they do not like to change frequently the international keyboard into the Romanian one. That is why most people do not use the letters with Romanian diacritics in the chat or Messenger communications. Therefore, many words become homograph and give rise to confusion, especially for the POS-tagger.

These confusions are funny or ridiculous, but this fact does not disturb the communicators; on the contrary, they often intend to play or to amuse themselves. For example:

– the word "putin" is ambiguous, it is an adverb, with the correct orthography "puțin" (a little). The POS tagger interprets it as a proper noun, Putin, the Russian president;
– The word "manca" is a main verb imperfective indicative third person singular, with the correct orthography "mânca", (to eat) but the POS tagger associates the form "manca" with the lemma "mancă" (nurse) and interprets it as a noun with the POS tag "Ncfsry", "Noun common feminine singular direct + definiteness" which determines the syntactic parser to assign it the dependencies of a noun and not the syntactic dependencies of a verb;
– "pana" is a related word, with the correct orthography "până" (until), with the POS tag "Sp" = "Adposition preposition simple accusative" and the POS-tagger associates lemma "pană" (plume) causing the selection by the parser of noun dependencies;

Other particular orthography is the lack of capitalization, which makes it impossible to distinguish from the proper nouns and the common ones. More difficult to understand are abbreviated words in many non-standardized ways.

Examples:

– nik is an abbreviated word which can be "nimic" (nothing) or "Nicolae" (Nicholas);
– k is a letter which can mean "ca" (like), adverb of comparison, or "că" (that), subordinate conjunction;
– fk can be read as "face" (does), verb main indicative present third singular, or "facă" (to do), verb main subjunctive present third.
– Pr, pt, ptr mean "pentru" (for), apposition preposition simple accusative.

We must remark that all these graphemes are easy to understand in the context, i.e. the communicative purpose is reached. The procedure to obtain such as abbreviations are: the vowel sounds are omitted because the consonant sounds transmit more information, or the ending of the word is omitted, because the information is located rather at its beginning.

To build a POS-tagger which will be able to correctly interpret all these non-standardized phenomena, we must introduce in its vocabulary the words without diacritics and also all the abbreviated words as lexical variants of words correctly spelled. It would be needed also a lot of contextual disambiguation rules for all resulting homographs. It would be more difficult to make the program read the information of consonants rather than of vowels or of the beginning of words rather than of their end.

Another problem is the lack of punctuation. As we stated in the introduction, we used of a minimal normalization (although we disapprove this method) by introducing periods in the texts extracted from the chat. The lack of commas makes the syntactic parsing more difficult, because commas separate dependencies blocks, avoiding the subordination of words at the other head than they should be. Commas make also a difference between the core dependencies and the facultative modifiers, and also it isolates the words or the groups of words which are not part of the main structure.

4.2 Pragmatic, Lexical and Stylistic Particularities

The purpose of this creative communication is to accurately convey to receivers the transmitter's states, feelings, ideas. Thereupon, the deviations from the rules of any type of standardized language are allowed if they do not disturb the communication. The social media language is free, without constraints. These deviations can be caused by haste or by comfort, may be unintentional and unnoticed. But we are particularly interested of the intentional deviations of the standardized rules that have stylistically conveyed an attitude: playful, ironic, parody or protest. If "aeriaesc" = "aerisesc" (I aerate the room) is an unintentional mistake, there are other graphical forms that are intentional stylistic deviations.

Examples:

- "aurzi?" = "auzi?" (do you hear?) is a contamination with the antonymous "surzi" (who cannot hear);
- "se întâmpulă" = "se întâmplă" (it happens) is a quite unobserved porno allusion of "pulă" (penis);
- "instifut" = "institut" (institute) in some way, it is a contamination with "fut" (fuck);
- "cerşetare" = "cercetare" (research) contaminated with "cerşi, cerşetor" (panhandle, beggar).

The other are clearly means of expression for the irreverent attitude towards the institution of research in question and towards the modality in which the Romanian research is financed.

The chat may contain also foreign words, in correct or incorrect form: "bonsoar" (Fr. bonsoir = good evening), "prego" (It. prego = please) "forghes mi nacht" (Germ. vergiss mich nicht = forget me not) being also contaminated with Germ. "Nacht" (night), etc.

The playful character appear as a continuous selection of styles, a sudden transition from one to the other. "cheatră" (stone) "tălică" (you) are regional terms, "nu sciu" is an archaic phonetic form (I don't know), the following are barbarisms, i.e. foreign words unnecessary, with wrong forms: "mă rancontrez" (I meet), "săptămâna futură" (the next week), "monșerule" (mon cher = darling).

Similarly important is the lexical inventiveness, i.e. many personal lexical creations that appear in social media communication.

Examples:

- "electoralnică" is a contamination between "electoral" and "jalnic" (lamentable);
- The tendency is to debunk, vile, coarsen the false values established.
- "bursă de ieșit afară, – bursă purgativă." (a scholarship to come out – to go abroad, purgative scholarship).
- The term "maestro" is commented in ironic style, associated with "meșter" (master) and "măiastră" (charmed, magic) the Grand Master of feminine gender becomes "mare măiastră" (great charmed) and master of ceremonies becomes mortician, person who "trage la pompe funegre" (Shoot to the funeral pumps). Another stylistic mean is the contamination of word "funeral" with the word "negru" (black).

In such cases, the POS-tagger will give errors, but fortunately, the social media users are not all so inventive creators of language. The possibilities for interpreting and commenting on the specifics of this means of communication cannot be exhausted in a single article. All the lexical inventions funded in our corpus inventions may migrate into the standard language; and therefore, it is a good initiative to study their attestations and their frequency for to registering them in the POS-tagger dictionary, with their semantic and formal deviations. We can also introduce in the parser a useful rule such as the tolerance of one or two character difference between lemmas of its lexicon and the word-text; it is required by the regional variants and the creative variants of the social media texts.

Another way to pragmatically interpret the social media texts is to observe in which modality the communicative roles and the relations with the communication situation are established. Unlike the dialogue in familiar style, where people are face to face, the chat is written, and the communicators are remote, but they make use of different conventions to establish the same kind of relationship as in a face to face conversation. For example, there are a lot of banal or creative modalities to fulfill the pragmatic functions of discourse:

- to establish the contact and begin a conversation between the transmitter and *the receiver: "Hei!" (Hey!), "Hey, ce faci?" (Hi! How are you?), "Ești acolo?" (Are you here?) "Ascultă aici" (listen here!)*

The same function can be fulfilled by creative means: *"săru-mânurile pă obraz!" (hands kiss on the cheek!)*

- to verify the existence of the link (of the canal to convey the information intended): receiver: *"Auzi?" (do you hear me?) "Ești on line?" or "Ești acolo?" (you are online? you there?)*
- to verify the communication code: *"înțelegi?" (do you understand?)*
- to ask or claim the role of the transmitter: *"să-ți spun ceva în privat" (I will tell you something private)*
- to keep the transmitter role, by proposing topics for discussion: *"să-ți spun una" (to tell a good one)*

– to pass of the other the transmitter role (to ask information): *"tu ce zici de?" (what do you say to that?)*
– to reject the proposed theme of the conversation:
  • *Tu îţi aminteşti de unul, Florin Toma? (You remember the one, Florin Toma?)*
  • (After more replies) *Dar de Toma, ce zici? (What about Thomas, eh?)*
  • *care, necredinciosul? (Which one, the unbeliever?)*
  • *Nu, Florin! (No, Florin!)*

The interlocutor pretends to forget whom a reply refers to, which wants to express the idea that the person does not deserve attention, or there are another reason for which this person refuse the theme proposed and does not offer the information asked.

4.3   Discourse Analysis and Interwove Topics of Conversation

The ludic (or the playful) character of this kind of communication is manifested also at the discursive level. The social media language is a conglomeration of styles. The participants of the communicative exchange adopted in turn various roles, talking like some fictional characters in the books of their admired authors, Shalom Aleihem or Caragiale, both humorists. In the following examples, the first communicator talks like a character from Shalom Aleihen, a Romanian Jewish person, and the conclusion of the conversation is line from Caragiale:

  • sentence id. 47–48: *Şi şe faşeţi, oi, oi, oi, cu atâtea? Miliţie, pompieri, vidanjă? And what you do, oh, oh, oh, with so many people? Police, fire, sewage truck?*
  • sentence id. 65: Bobocule, *cum le ştii tălică pe toate!* (My bourgeon, how well do you know them all!)

The intertwining of themes is a well-studied phenomenon in chat conversations. See, for instance, the paper "The Right Frontier Constraint Holds Unconditionally" by Dan Cristea [2]. The rule of the new structures attached only or habitually in the right part of the tree is similar of the syntactic rule of the position of the subject, because the subject represents the information known and the predicate express the information gain. In English, these rules are strict. If Chomsky's Universal Grammar had been written knowing not only English but also Latin, we would have had a different understanding of the formalized morphology and syntax. Lucien Tesnière, the first researcher to note the existence of infinitive and gerundive sentences, does it because he was a connoisseur of Latin.

For example, we notice that the subject is a linear function with two values, the position in the string/vs/the morphological marks, each member of the disjunction are measured by a parameter vernacular or a filter: if the morphological mark of the word is invalid, the system must go to word's position in the sentence. It means that in English the inflexion of words is reduced and only the position in the string is relevant for their function, there is a right frontier constraint (RFC) that holds unconditionally, i.e. there is a strong constraint the word order to be respected, always the new information is added to the right border of the discourse tree, but only in English or in the other languages that have a reduced morphology. In the cited paper, the chat text is an example for the deviation from the right frontier constraint.

However, in the languages with a rich morphology, such as Romanian, the position in the string is not significant for the word function; the right frontier constraint does not exist for Romanian. The noun with definite article has a lot of flexional forms, and the subject can be positioned at the end of the string, the direct object can precede it. The verb has also a bigger number of flexional forms, that is why the subject, the theme, can be elliptical in the string; if it is a personal pronoun, the form of the verb take the information about the person and the number of the agent.

That are causes for the parser mistakes, it is searching for the subject first, for the predicate in second place and for the direct objet in the third, but the examples in our Treebank have not this structure. We need a rule based or a hybrid parser to introduce in the program syntactic rules and semantic information, such as: for a certain verb, the agent can be only a human and the patient can be only an object, etc.

For the same reason, we do not accept the fourth axiom of the dependency grammar, which prohibits edges intersecting shaft: If A depends on B, and C is an element between A and B (according to the linear order of the sentence), then C depends directly either on A or on B, or on another item found between A and B [4].

There is an isomorphism between the syntactic layer, the semantic layer and the discourse layer. The patient can precede the agent, and the theme, the information known, can be placed after the rheme, containing the information gain. Consequently, the transmitter can construct his or her message by adding a sub-tree in any node of the tree, in the right side or in the left side.

Another particularity of this style is caused by the written form. While one person is typing a reply, he or she cannot read what his/her interlocutor wrote, perhaps he/she opens another theme of conversation, and later he/she reads the answer of his/her interlocutor to the previous reply. Often the conversation is on several themes interweaved. The syntactic parser cannot analyze this kind of phenomena which do not appear in the sentence but between many sentences; we need a tool for the discourse analysis. We emphasize again that the communicative purpose is achieved, the interlocutors know perfectly well where every line is related, but if they pretend to ignore that, it will create humorous effects.

In some sort, this feature of the discursive organization, i.e. to interweave parts of two storylines unconnected one with another, shows that obviously these communicators are lonely people, who tell each other their own story without in fact relating to each

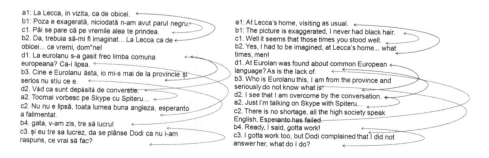

Fig. 1. Interwove themes of more interlocutors.

other. They do not want to know the history of other, the two partners refuse to quite their theme accepting the theme proposed by the interlocutor, the communication is difficult and perhaps it does not exist.

The situation of more than two participants is more complicated. In general, our corpus includes conversations between two parties, but conversations with several participants are current Social Media and should also be studied. An example, in which there are four interlocutors, has a most complicated structure of the relations between the replies:

In the case of more interlocutors, the interweave of themes has the character of the rumors. In the example in Fig. 2, there are four themes:

The first theme is the commented photo, the event in the picture and the color of the hair. This theme is not related with the others. The second theme is the Eurolan; someone asks the meaning of this is while another person is interested in the common European language. The second question is answered, the first receives no answer. The person interested by the European common language protests because he/she does not receive immediately an answer, and the person that had asked the first question introduces the fourth theme, the work; that being a hidden protest for the lack of the answer at her question.

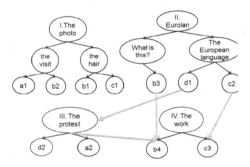

Fig. 2. The interwove themes of the conversation in Fig. 1.

The theme of the work is a pragmatic mark of the desire to end the discussion, and the last reply is an agreement with the end of the conversation, but, as an excuse to prolong it, it remains the necessity of an answer related to the common European language, because the interlocutor has protested for not having received an answer. There are hidden or explained relations between the II, III and IV themes, despite the fact that the three themes are marginal to the principal theme of the conversation, the picture. The first theme has no relation with the following themes. The old picture is in fact a pretext to exchange ideas with the persons in the picture, which are at great distances, in different countries.

5   Evaluation of UAIC POS-Tagger and of the Syntactic Parser Accuracy for the Social Media Corpus

A POS tagging evaluation has been conducted using the UAIC POS-tagger for Romanian [15]. The gold corpus for this evaluation consisted of only the first 200 sentences from the

Romanian social media corpus. These sentences have been manually corrected by a human annotator.

The social media corpus contains Romanian texts with and without diacritics. It is perfectly acceptable for this kind of communication to contain in one sentence both words which use diacritics and words which do not use diacritics, because the communicators pay no attention to this aspect. This slows down the automatic part of speech tagging. From a data processing point of view, Romanian texts that do not use diacritics have to be separated from Romanian texts that correctly make use of diacritics. This is due to the high frequency and importance of diacritics in Romanian.

To overcome this, a separate POS tagger has been trained on a mixed version of the Romanian corpus (with and without diacritics), while using a mixed morphologic dictionary. As expected, this so-called "mixed diacritics POS tagger" has lower performance than its precursor, which has been trained on the corpus with diacritics only. This is due to an increased level of morphological ambiguity.

The training corpus used for the diacritics-only POS tagger [15] consisted of the newspaper texts from NAACL 2003[1] plus 28,000 sentences extracted from JRC-ACQUIS[2], resulting in a total of 67,000 sentences. The gold corpus developed in the Multext-East project, Orwell's "1984", was used for evaluating the POS-tagger accuracy. But this POS tagger has a very slow accuracy for the social media sentences.

For training purposes, a "mixed diacritics" Romanian POS tagger, both the training corpus and the test corpus have doubled in size, by appending their stripped diacritics versions. To accommodate the modifications in the training and testing data, the morphologic dictionary has been injected with the stripped diacritics version of its entries and thus increasing both its size and morphologic ambiguity.

The POS tagger also uses a set of Graphical Grammar Studio correction rules [16], which also required modifications. In this way, the accuracy for the social media text has significantly increased, although it is not yet satisfactory. We intend to train the parser with 2,576 sentences of the chat sub-corpus and to increase the number of the rules. For example, functional words, such as conjunctions, prepositions, relative pronouns and adverbs are less numerous and most frequent and we can make rules for these words in their versions with and without diacritics. A tool for the disambiguation is also necessary.

Table 1 depicts the POS tagging deficiencies when tagging mixed Romanian texts having letters with or without diacritics in Social Media communication. In Social Media, the letters with diacritics are rarely found, but they exist sometime, so the POS-tagger must recognize the two orthographies. As can be seen in our examples above, there are a lot of ambiguities appearing in the texts containing both orthographies, which increase the parsing difficulties of the Romanian language.

---

[1] A parallel corpus for Romanian-English created at the HLT/NAACL 2003 workshop, titled "Building and Using Parallel Texts: Data Driven Machine Translation and Beyond".
[2] JRC-ACQUIS is the largest parallel corpus. It is composed of lows for the EU Member States, since 1958 till present, translated and aligned for 23 languages.

Table 1. Evaluation method of UAIC Romanian POS tagger for standardized texts

Diacritics only RO-POS tagger, eval. on 1984	Diacritics only RO-POS tagger, eval. on 200 sent. from the corpus	Mixed diacritics RO-POS tagger, evaluated on mixed 1984	Mixed diacritics RO-POS tagger, evaluated on the first 200 sentences from the chat corpus
97.03 %	54.33 %	94.38 %	74.53 %

Experiments regarding syntactic parsing have also been conducted. The UAIC Romanian syntactic parser is based on the Malt parser tool [12]. The UAIC parser model and the training corpus developed together in a series of at least 3–4 years, as a result of many bootstrapping cycles, and this process will be continued. Today, the Romanian UAIC syntactic parser is available as an online tool[3], both as a web application and a web service.

Table 2. Evaluation of UAIC Romanian syntactic parser for standardized texts

Metrics	Both attachment	Head attachment	Label attachment
Standard type of texts	78.04 %	84.25 %	83.57 %
Social media text	58.31 %	71.74 %	66.08 %

These results contain the precision of the head detection (column 3), the precision of the label (dependency relation) attachment (column 3) and the precision of both these parameters (column 2). An important aspect of the results in Table 2 is that the training data does not contain any social media texts. A separate experiment, which also includes social media text in the training data, has been conducted. For this experiment, all the texts from the UAIC Romanian Dependency Treebank and the social media texts presented in this paper have been brought together in a 10-fold scheme. The results presented in Table 3 clearly demonstrate the importance of training data for social media syntactic analysis. The overall score drops slightly. This is explained by the many linguistic particularities of social media texts. But when evaluated only on the social media slice of the test corpus, the improvements are clearly visible:

Table 3. Evaluation of UAIC Romanian syntactic parser trained on both standardized text and social media texts

Metrics	Both attachment	Head attachment	Label attachment
Standard type of texts	77 %	83.65 %	82.99 %
Social media text	66.01 %	78.48 %	72.21 %

These results do not contain the percent of the "good trees" entirely correctly parsed. This percent is yet very slow. The serious errors come from the undetected root and upper levels of the tree. Apart from increasing the size of the training corpus, we consider using another parser, based on both statistics and rule based.

---

[3] UAIC Romanian dependency parser http://nlptools.infoiasi.ro/WebFdgRo/.

In the corpus of 1 000 sentences, in order to enable the comparison and the evaluation of POS-tagger, 10.05 % of sentences have been removed, the changes of the split have been manually made, by adding or removing portions of the text, in the.xml format, since they cannot be aligned with the automatically annotated variant, having a different number of units, for the evaluation of the accuracy. This percent represents the splitting errors (manually revised) made by the POS-tagger processing the euro media texts.

By comparing the accuracy of the POS-tagger and the accuracy of the parser for each item, we found that 43.13 % of labeling errors are superposed on POS-tagging errors, i.e. they are caused by it. This result leads us to conclude that we should focus our attention in the first stage of this research on the growth of POS-tagger accuracy for the social media texts, without neglecting the syntactic parser trained on social media.

We need a strong procedure for the disambiguation of similar form words caused by the written form without Romanian letters with diacritics. The gold corpus for the POS-tagger training on Social Media texts will be increased in the future experiments.

6 Conclusions and Future Works

The social media is an important means of communication in the contemporary society, without which we cannot imagine the relations between individuals now and in the future. An increasing volume of such texts can be collected on the web. That is why social media texts must be included in the corpora for each natural language. But the automatic annotation of this kind of texts is very difficult [18].

The tools for natural language processing are built for standardized communications, and the social media is the non-standardized empire of the absolute freedom of communication. Supporting only the pragmatic rules, that message is constrained only by the need to optimize the transmission of the content intended to the recipient. The transmitter uses the most creative stylistic means to achieve the communicative purpose. The standard rules of the language, built to facilitate the communication, accepted by all the speakers, start to trivialize to such an extent in use, that they cannot express the communicative and precise intentions anymore; they are suspended by the endowed communicators with a high creativity.

Seeing that we cannot give up the study of this stylistic variant of the language, what remains to be done is to formulate as many rules as necessary to optimize the running of the NLP tools on this type of messages.

Acknowledgement. We are grateful to members of the NLP group in our faculty for suggesting us such an interesting research topic.

References

1. Avontuur, T., Balemans, I., Elshof, L., van Noord, N., van Zaanen, M.: Developing a part of speech tagger for Dutch tweets. Comput. Linguist. Neth. J. **2**, 34–51 (2012)
2. Cristea, D.: The right frontier constraint holds unconditionally. In: Proceedings of the Multidisciplinary Approaches to Discourse 2005 (MAD 2005), Chorin/Berlin, Germany (2005)
3. Dent, K., Alto, P., Diep, F.: Parsing the twitterverse. Sci. Am. **305**, 22 (2011)
4. Darling, W., Paul, M., Song, F.: Unsupervised part-of-speech tagging in noisy and esoteric domains with a syntactic-semantic Bayesian HMM. In: Proceedings of the Conference of the European Chapter of the Association for Computational Linguistics, pp. 1–9. ACL (2012)
5. Derczynski, L., Maynard, D., Aswani, N., Bontcheva, K.: Microblog-genre noise and impact on semantic annotation accuracy. In: Proceedings of the 24th ACM Conference on Hypertext and Social Media. ACM (2013)
6. Foster, J., Cetinoglu, O., Wagner, J., Le Roux, J., Hogan, S., Nivre, J., Hogan, D., van Genabith, J.: POS tagging and parsing the twitterverse. In: Proceedings of the AAAI Workshop on Analyzing Microtext (2011)
7. Gadde, P., Subramaniam, L., Faruquie, T.: Adapting a WSJ trained part-of-speech tagger to noisy text: preliminary results. In: Proceedings of the Joint Workshop on Multilingual OCR and Analytics for Noisy Unstructured Text Data, pp. 5:1–5:8. ACM (2011)
8. Gimpel, K., Schneider, N., O'Connor, B., Das, D., Mills, D., Eisenstein, J., Heilman, M., Yogatama, D., Flanigan, J., Smith, N.: Part-of-speech tagging for twitter: annotation, features, and experiments. In: Proceedings of the 49th Annual Meeting of the Association for Computational Linguistics: Human Language Technologies, pp. 42–47. ACL (2011)
9. Liu, F., Weng, F., Jiang, X.: A broadcoverage normalization system for social media language. In: Proceedings of the 50th Annual Meeting of the Association for Computational Linguistics (2012)
10. Neunerdt, M., Trevisan, B., Reyer, M., Mathar, R.: Part-of-speech tagging for social media texts. In: Gurevych, I., Biemann, C., Zesch, T. (eds.) GSCL 2013. LNCS, vol. 8105, pp. 139–150. Springer, Heidelberg (2013)
11. Neunerdt, M., Reyer, M., Mathar, R.: A POS tagger for social media texts trained on web comments. Polibits **48**, 59–66 (2013)
12. Nivre, J., Hall, J., Nilsson, J.: MaltParser: a data-driven parser-generator for dependency parsing. In: Proceedings of the Fifth International Conference on Language Resources and Evaluation (LREC), May 24-26, 2006, Genoa, Italy, pp. 2216–2219 (2006)
13. Owoputi, O., O'Connor, B., Dyer, C., Gimpel, K., Schneider, N.: Part-of-Speech Tagging for Twitter: Word Clusters and Other Advances. Technical Report CMU-ML-12-107, Machine Learning Department, Carnegie Mellon University (2012)
14. Robinson, J.J.: Dependency structures and transformational rules. Language **46**, 259–285 (1970)
15. Simionescu, R.: Hybrid POS tagger. In: The Workshop on Language Resources and Tools in Industrial Applications, Eurolan 2011 summer school (2011)
16. Simionescu, R.: Graphical grammar studio as a constraint grammar solution for part of speech tagging. In: The Conference on Linguistic Resources and Instruments for Romanian Language Processing (2011)

17. Singha, K.R., Purkayastha, B.S., Singha, K.D.: Part of speech tagging in manipuri: a rule-based approach. Int. J. Comput. Appl. (0975 – 8887) **51**(14), 31–36 (2012)
18. Toutanova, D., Klein, C., Manning, C., Singer, Y.: Feature-rich part-of-speech tagging with a cyclic dependency network. In: Proceedings of the Conference of the North American Chapter of the Association for Computational Linguistics on Human Language Technology, pp. 173–180. ACL (2003)

Author Index

Printed in the United States
By Bookmasters